Assessing Pittsburgh's Science- and Technology-Focused Workforce Ecosystem

MELANIE A. ZABER, LINNEA WARREN MAY, TOBIAS SYTSMA,
BRIAN PHILLIPS, STEPHANIE J. WALSH, ROSEMARY LI,
ELIZABETH D. STEINER, JEFFREY B. WENGER,
ÉDER SOUSA, JESSICA ARANA

Supported by the Richard King Mellon Foundation, the Heinz Endowments,
the Henry L. Hillman Foundation, and the Claude Worthington Benedum Foundation

EDUCATION AND LABOR

For more information on this publication, visit **www.rand.org/t/RRA1882-1**.

About RAND

The RAND Corporation is a research organization that develops solutions to public policy challenges to help make communities throughout the world safer and more secure, healthier and more prosperous. RAND is nonprofit, nonpartisan, and committed to the public interest. To learn more about RAND, visit www.rand.org.

Research Integrity

Our mission to help improve policy and decisionmaking through research and analysis is enabled through our core values of quality and objectivity and our unwavering commitment to the highest level of integrity and ethical behavior. To help ensure our research and analysis are rigorous, objective, and nonpartisan, we subject our research publications to a robust and exacting quality-assurance process; avoid both the appearance and reality of financial and other conflicts of interest through staff training, project screening, and a policy of mandatory disclosure; and pursue transparency in our research engagements through our commitment to the open publication of our research findings and recommendations, disclosure of the source of funding of published research, and policies to ensure intellectual independence. For more information, visit www.rand.org/about/research-integrity.

RAND's publications do not necessarily reflect the opinions of its research clients and sponsors.

Published by the RAND Corporation, Santa Monica, Calif.

© 2023 RAND Corporation

RAND® is a registered trademark.

Library of Congress Cataloging-in-Publication Data is available for this publication.

ISBN: 978-1-9774-1048-1

Cover credits:
Map: FrankRamspott/Getty Images
Left to right: poba/Getty Images; TommL/Getty Images; f11photo/Getty Images; FG Trade/Getty Images

Cover composition and figures: Jessica Arana

About This Report

In this report, we discuss the findings from a quantitative and qualitative assessment of the science- and technology-focused (STF) workforce ecosystem in the seven-county Pittsburgh region. This assessment focuses on the demand for and supply of STF workers, training pipelines, anticipated growth, unique regional assets, and liabilities for supporting Pittsburgh's STF ecosystem. The findings and recommendations are intended for local education and training institutions, workforce centers, employers, economic development organizations, policymakers, funders, and residents of the Pittsburgh region.

A separate document containing seven appendixes is available at www.rand.org/t/RRA1882-1, and an accompanying tool is available at www.rand.org/t/TLA1882-1.

RAND Lowy Family Middle-Class Pathways Center

This research was conducted within the RAND Lowy Family Middle-Class Pathways Center. The center aims to identify ways to create and sustain middle-class employment in the face of rapidly changing labor-market conditions. The center is part of RAND Education and Labor, a division of the RAND Corporation that conducts research on early childhood through postsecondary education programs, workforce development, and programs and policies affecting workers, entrepreneurship, and financial literacy and decisionmaking.

For more information about the RAND Lowy Family Middle-Class Pathways Center, visit www.rand.org/mcpc. For more information on RAND Education and Labor, visit www.rand.org/education-and-labor. Questions about this report should be directed to Melanie Zaber at mzaber@rand.org.

Funding

This study was supported by the Richard King Mellon Foundation, the Heinz Endowments, the Henry L. Hillman Foundation, and the Claude Worthington Benedum Foundation.

Acknowledgments

The authors would like to thank Matthew Crespi and the Seldin/Haring-Smith Foundation for sharing their public transit map with our research team. Asya Spears provided valuable research support. We are grateful to our three reviewers, Sabina Dietrick, John Engberg, and Catherine Augustine, for providing comments that improved the report. We would like to acknowledge the advice of local science and technology experts who provided guidance on the construction of our definition of science- and technology-focused jobs. We thank our focus group participants for lending additional context and perspective to the report's quantitative findings. Finally, we thank the program officers from the four sponsoring foundations for their participation in a workshop on project recommendations and their guidance throughout the project.

Summary

Background

Over the past decade, more than 10 billion dollars has been invested in Pittsburgh tech companies, with more than 3.5 billion invested in 2021 alone (Burkholder, 2022). More recently, tens of millions of dollars were invested in the Pitt BioForge Biomanufacturing Center that will soon be home to ElevateBio and other biotech companies (Conway, 2022). Such strong sectoral growth raises the following questions: What kinds of jobs are needed for and supported by this growth? and What investments should be made to continue propelling the region's science- and technology-focused (STF) sectors into the future? With this context in mind, RAND Corporation researchers set out to characterize the STF workforce ecosystem in the Pittsburgh region and suggest policy changes and investment opportunities to "future-proof" the ecosystem.

Project Goals and Approach

The focus of this project is the Pittsburgh region, the seven-county metropolitan statistical area (MSA) that includes the city of Pittsburgh. The goals of the project were to define STF occupations in a regionally relevant way, characterize the current state of the STF ecosystem, identify barriers and facilitators to participation in the STF ecosystem, and develop strategies to facilitate the STF ecosystem's continued growth. To achieve these goals, the research team used qualitative and quantitative methods. Qualitative data informed the scope and types of quantitative analyses conducted, and quantitative data analysis served both as a prompt for focus group discussion and to verify and qualify select perceptions raised. Additionally, the research team assessed the qualitative data for context and to provide potential explanations for patterns observed in quantitative data. Finally, the team selected Boston and Nashville as peer regions to further contextualize quantitative findings.

Key Findings

Pittsburgh has a sizable STF workforce. Approximately 18 percent of Pittsburgh's employment is in STF occupations, outpacing the national share (16 percent). Pittsburgh's STF workforce grew as a share of the overall labor force between 2015 and 2019. Still, there are indicators that even this size STF workforce may be insufficient for the region. The pace of STF growth suggests the need for more workers in the future, in contrast to the region's declining population. And focus group participants reported challenges in filling open positions, particularly mid- to senior-level managerial positions (a sentiment that was echoed in Tripp et al., 2021, a recent report on the region's robotics and autonomy cluster). Additionally, there is increasing national competition for STF workers as many regions are trying to position themselves as market leaders in high-tech industries (Tripp et al., 2021).

Most of the demographic trends in Pittsburgh's STF workforce are driven by regional demographics. Although Pittsburgh's STF workforce (and the region overall) became slightly more racially and ethnically diverse between 2015 and 2019, the region—and its STF workforce—has substantially less racial and ethnic diversity than peer regions and the United States overall. Many focus group participants noted difficulties recruiting and retaining workers of color. For instance, some participants perceived a lack of exposure to STF jobs for all workers and unclear pathways into and within the STF workforce for local workers of color. Other participants reported their perception that the competitive national marketplace for diverse

STF talent made attracting workers from underrepresented groups to Pittsburgh more challenging. Like in the peer regions of Boston and Nashville, 44 percent of Pittsburgh's STF workers are women. However, Pittsburgh's STF workforce (like the local population) is older than those in peer regions, with an average age of 43. Over the past decade, the Pittsburgh region's population has shrunk and aged, with particularly acute declines in residents under the age of 25 (outpacing slight declines nationally).

Pittsburgh's STF workforce has a smaller share of foreign-born workers—and workers born out of state—than the STF workforces in Boston and Nashville. However, Pittsburgh STF workers are more likely to be foreign-born than workers in the region's overall labor force (which is not true in the peer comparison regions), suggesting that STF employers may use different hiring strategies. A lack of in-migration to the region (from both outside the state and outside the United States), coupled with population losses, threatens the region's future ability to supply a workforce for growing companies.

The counties surrounding Allegheny are key contributors to the STF ecosystem, but their residents may face some barriers. Although the majority of the region's STF workers live within Allegheny County, 42 percent of STF workers live in the counties surrounding Allegheny. Allegheny and Butler counties both contribute a disproportionate share of the region's STF workforce. In addition to these "people assets," focus group participants perceived that the assets of the surrounding counties, such as "shovel-ready" potential developments, are currently underused and could enhance the counties' contributions to the STF ecosystem.

Currently, STF workers who live in the surrounding counties may have to overcome geographic barriers to engage with the STF ecosystem. Fifty-eight percent of STF workers who live in the surrounding counties work outside their home county or remotely, compared with just 22 percent of Allegheny County STF workers. STF education and training programs are available at most credential levels in most counties in the region, although Allegheny County is home to about 60 percent of the region's STF education providers. A lack of transportation options may be a particular challenge for STF learners and workers in surrounding counties—education and training programs are less likely to be located close to a bus stop, and residents' homes may be similarly far from transit. Remote learning options are available, but associate degrees and sub-baccalaureate certificates in STF fields tend to exclusively offer in-person classroom learning and hands-on education. The region's STF wages, which are low relative to national levels, are relatively consistent across Pittsburgh MSA counties, which might help offset transportation costs.

Sub-baccalaureate STF employment plays a vital role in STF sectors and the broader regional economy. The region's STF ecosystem employs workers from various educational backgrounds and creates economic activity throughout the region. Thirty percent of Pittsburgh's STF employment is in jobs that do not require a bachelor's degree, and five of the top ten STF occupations projected to have the most openings over the next decade do not require a four-year degree. The region's sub-baccalaureate STF jobs also create employment regionally through workers spending their incomes and through business supply-chain interactions. The average Pittsburgh STF job creates one additional job somewhere in the region, which is 30 percent more local employment than the average non-STF job creates. Pittsburgh's sub-baccalaureate STF jobs generate an outsized "ripple effect" in employment compared with Boston and Nashville.

The region's large and highly concentrated health sector may limit spillover benefits and suppress wages. A large share of Pittsburgh's STF employment is in the health sector, though Pittsburgh's industrial past also contributes to the region's relatively large share of STF workers in technician and production-related industries. Production-related employment generates large spillover benefits for the region; health jobs generate comparatively less additional employment. Pittsburgh's health sector is more concentrated, with fewer employers for workers to choose from, than the health sectors of other regions. Pittsburgh's health workers make about 10 percent less than the national average for their occupations, which is not offset by a lower cost of living.

STF wages in the region allow homeownership, but still fall behind those in other regions, even after adjusting for cost of living. Pittsburgh STF jobs pay enough to buy a house as a single worker and pay enough to support a family in a dual-income household. Pittsburgh outperforms other regions in this area: Ninety-five percent of STF workers work in an occupation where the average salary makes the median-priced home affordable for an individual earner, compared with 63 percent of STF workers in the United States and just 30 percent in Boston. With housing becoming increasingly unaffordable in many STF hubs (Anthony, 2022), Pittsburgh's stock of affordable housing could be a considerable asset for drawing in younger workers looking to build equity in the housing market. However, the perceived affordability of Pittsburgh's housing is largely attributable to the region's stock of inexpensive housing rather than to the region's wages. After adjusting for local cost of living, Pittsburgh STF workers make an average of 8 percent less than the national average for their occupation. Additionally, the region's asset of an affordable housing market appears to be less accessible to workers of color than in comparison regions, with lower rates of homeownership relative to Pittsburgh's White residents and relative to workers of color in other regions.

Data challenges hinder communication of career pathways and verification of student outcomes. Pennsylvania lags peer states, such as Kentucky, West Virginia, and Texas, in tracking student data from kindergarten through grade 12 (K–12) into postsecondary education and into the workforce. This makes it challenging to credibly communicate the likely outcomes after pursuing a given education or training program, which may dissuade potential workers from pursuing STF training for an uncertain payoff. Focus group participants shared that individual education and training program providers struggle to keep track of where participants move or are employed when they leave or graduate.

The long-term regional implications of remote work in STF jobs are unclear. Pittsburgh STF jobs are less teleworkable relative to those in comparable cities because of the region's higher share of health and technician/production employment, both of which are primarily place-based. Still, many of the region's STF jobs are teleworkable. As a result, Pittsburgh STF employers—who tend to be smaller than their counterparts in Boston and Nashville—may see increased competition nationally because of the growth of remote work. However, remote work may offer local employers the opportunity to hire from a wider labor market, provided that local salaries can compete nationally. Ultimately, how Pittsburgh fares in a remote work world depends on where the remote work wage settles relative to the region's current wage.

Regional educational providers are well positioned to prepare students for STF skills that will grow in demand. Human resource skills, physical skills, project management skills, and coding skills will likely play a vital role in the Pittsburgh STF ecosystem in the future. Machine learning skills and machinery skills are also growing in demand across multiple occupation groups. Our inventory of STF education and training programs suggests that the region's existing educational institutions and programs could support skill-building through short- and longer-term programs in these areas. However, infrastructure alone is not enough—the region also needs a sufficient supply of students informed about, interested in, and capable of completing these programs. Regular collaboration between regional employers and educational providers can help ensure that programs are developed and refined to meet evolving skill needs, while a focus on reducing barriers to program access and completion, including cost, transit accessibility, and a need for wraparound services (e.g., career counseling, child care vouchers, health support), can facilitate program participation.

Recommendations

Overall, we found that Pittsburgh has a sizable share of STF employment relative to the United States and to Nashville, one of two chosen comparison cities. However, additional investments and changes to policy can safeguard the region's strengths and support Pittsburgh as a flourishing science and technology hub. We briefly discuss each study recommendation, along with potential elements of implementation. We recognize that there are many extant efforts to address these challenges and our discussion of remaining challenges in the region can help further support these and other efforts in the region.

Create the Market Conditions Needed to Expand the STF Workforce

In light of the region's aging and declining population, STF economic growth may not be sustainable without expanding the STF workforce. Investing in policies that improve the STF ecosystem's national competitiveness, and thus increase wages, could expand the region's STF workforce. For instance, policies that foster competition by reducing barriers to entry and encouraging the development of new firms could improve the productivity of the STF ecosystem and drive economic growth. Policies that reduce the risk that employers face when investing in workers could help retain existing workers and build human capital in critical STF skills. Policies aimed at closing information gaps between workers and employers can help "future-proof" the region's STF pipeline by ensuring that workers are providing the skills that local employers need to be competitive with other regions. Aligning these sorts of policies with the region's existing strengths in education and training institutions, local energy production, and community amenities could help push regional wages up to a more nationally competitive level. Such policies might ensure that the region continues to attract and retain STF employers and STF workers from both within and outside the region, even as other regions position themselves as national STF hubs. In particular, such policies could help the region attract and retain young workers, who are critical to sustaining the future growth of the STF ecosystem.

Support and Engage Communities of Color and Other Locally Underrepresented Groups to Expand the Local STF Workforce and Meet the Ecosystem's Evolving Needs

Prior research shows that a diverse workforce increases innovation, improves firm profitability, and supports future recruiting. Moreover, many national employers have explicit diversity, equity, and inclusion goals that require a diverse workforce, such that they might be reluctant to open offices in regions without diverse talent pools to draw on (Shook, 2021). In the Pittsburgh region, fielding a diverse STF workforce will mean engaging communities of color, including immigrants; ensuring pathways to education and the workforce for people living outside Allegheny and Butler Counties; and retaining or attracting more workers under the age of 25.[1] However, these demographic groups are in short supply locally, meaning that the region will need to either attract these groups from other regions or better support local residents from these groups in their potential pursuit of STF careers. Addressing the racial disparities in public-facing systems (such as health care and lending) to demonstrate the region's commitment to equity and attract residents from outside the region, building community among learners of color to support their long-term retention in the region, and researching—and investing in—the community attributes valued by learners and workers of color can shift the region's demographic composition. Example approaches include using cohort-based models to establish scholarships for learners of color pursuing technical training or advanced degrees or establishing a local satellite campus for minority-serving institutions to attract a critical mass of diverse talent to the region. These efforts can help propel STF workforce growth and attract and support those employers who are committed to employing a demographically diverse workforce, growing the STF ecosystem as well. Finally, the informational improvements outlined in the next recommendation (on career

---

[1] Pittsburgh's STF workforce leans male (56 percent), but this is comparable to peer communities and thus does not put the region at a competitive disadvantage for attracting STF firms. However, growing gender diversity in the STF workforce could further enhance the productivity of the region's STF ecosystem.

pathways) can help highlight the viability of and rewards from pursuing STF career pathways locally. More-complete information can help engage populations with less exposure to STF careers, such as learners of color, learners from rural communities, and learners of lower socioeconomic status.

Build Regionally Relevant, Data-Backed Career Pathways

Career pathways are an organized set of jobs or occupations that map out a career trajectory through experience and education. To be most informative to potential students and workers, these pathways need to be specific to the regional labor market and education offerings and backed by evidence of local residents having successfully navigated those pathways. In Pittsburgh, data challenges hinder communication of career pathways and verification of student outcomes—the state's longitudinal tracking of student outcomes lags that of peer states. Instating a network of career pathways requires tracking learner outcomes across a broad variety of education and training programs (including boot camps, short-term credential programs, and colleges and universities), validating these career pathways with employers, and disseminating information through trusted community-based organizations. This process can help potential STF workers identify realistic career pathways and expand the number of appropriately skilled STF workers to support future growth. Additionally, collecting data on existing career pathways and program outcomes could provide scaffolding for future researchers to evaluate education and workforce interventions with longitudinal, individual-level data.

Craft and Implement a Regional STF Strategy

Finally, Pittsburgh has a variety of regional assets that can help drive growth in the STF workforce ecosystem, such as its large and lower-cost housing stock, world-class universities, and land for physical growth. Crafting and implementing a truly regional and multisector STF strategy can help the region better use resources outside Allegheny County (such as shovel-ready spaces for economic development) and create opportunities for ecosystem growth. Although some regional partnerships currently exist, they are generally focused on one or two sectors. Employers, workers, education and training providers, and community development organizations working together with trusted local leaders to craft and implement a regional STF strategy would allow the region to benefit from scale and prioritize sectoral diversification, improving the resilience of the STF ecosystem.

Contents

Figures and Tables

Figures

Tables

CHAPTER 1

Introduction

Pittsburgh today is a midsize city, outside the top 50 by population in the United States. But in its heyday, in 1950, Pittsburgh was the 12th largest city in the United States, just behind Boston and San Francisco (Gibson, 1998). During the postwar period, Pittsburgh and its suburbs underwent a renaissance, developing into both a corporate center and "the world's leading center for the development of nuclear reactors for naval vessels and power plants" (Vitale, 2021). Pittsburgh's manufacturing economy peaked in the 1950s; by the 1960s, Pittsburgh was home to the third largest concentration of Fortune 500 companies in the United States (Vitale, 2021).

In the postwar era, the Pittsburgh region was still a stronghold in manufacturing—particularly in iron, glass, and steel production—but in the 1980s, steel mills laid off hundreds of thousands of workers across the United States, with devastating impacts on the region. The cascading impacts from lost steel manufacturing jobs led to the loss of more than 8 percent of the region's jobs (Venkatu, 2018), and the resultant drop in disposable income fueled further business closures in the region (Hoerr, 1988). Manufacturing, once a sector in which the region outpaced the United States overall, declined in Pittsburgh throughout the 1980s so much so that the region fell below the national average in manufacturing employment share (Venkatu, 2018). In the 1980s and 1990s, the region shifted its industrial focus, leveraging the prestige of local universities, along with strong employment in health and banking, and began investing in technology-focused economic development (e.g., the construction of the Pittsburgh Technology Center). Although Pittsburgh's population has shrunk relative to the populations of Boston and San Francisco since the 1950s, these cities' names are again found together on lists of top U.S. cities, now for their robust existing and emerging tech sectors. The Pittsburgh of the postmanufacturing era is often cited in discussions around robotics, artificial intelligence, and automation. In 2016, a Brookings report found that the region's per capita university research and development (R&D) spending was nearly two and a half times the national average (Andes et al., 2017). The authors concluded that Pittsburgh was at a crossroads: It had the potential to become a world-class innovation region and the potential to remain a "could-have-been."

Pittsburgh is also frequently the recipient of accolades, such as being ranked one of the most livable cities in the United States (KDKA-TV News Staff, 2021) and at the top of a global Housing Affordability Index in 2022 (Cox and Lucyshyn, 2022). Housing affordability may help the region attract workers from outside the state, particularly as remote work proliferates. The region boasts a growing technology sector (with strengths in robotics and advanced manufacturing) featuring a mix of locally grown companies (e.g., Duolingo) and satellite offices of larger corporations (e.g., Google, Apple, Facebook). Seattle-based tech news website GeekWire performed its own "HQ2" search alongside Amazon and chose Pittsburgh as its winner (Nickelsburg, 2017).

However, five years after the Brookings report (Andes et al., 2017), it is unclear whether the potential future of Pittsburgh as a leading generator of innovative jobs is any closer to materializing. The region's startup leaders have repeatedly noted a lack of early-stage funding opportunities for entrepreneurs, which poses a barrier to translating R&D activity into economic growth (Tascarella, 2021; Tascarella, 2022). And in spite of the growing economic activity in the region, Pittsburgh's celebrated livability may not be experienced equally by all residents (Deitrick and Briem, 2021; Howell et al., 2019).

To understand the emerging scientific and technology sectors in Pittsburgh, their potential future, and the investments that can safeguard Pittsburgh from being a "could-have-been," a group of Pittsburgh-based foundations engaged the RAND Corporation to assess Pittsburgh's science- and technology-focused (STF) workforce ecosystem.[1]

Project Goals and Approach

Motivated by prior research looking at the future of work in the Pittsburgh region (Allegheny Conference on Community Development, 2017)[2] and the region's potential to become a leading innovation hub (Andes et al., 2017),[3] we assessed Pittsburgh's STF workforce ecosystem. The goals of the project were to

- define *STF occupations* in a way that acknowledges the contributions of Pittsburgh's large sub-baccalaureate workforce
- characterize the current state of the regional STF ecosystem, in terms of workers,[4] jobs, skills currently and likely to be in demand, education and training institutions, and other workforce stakeholders
- identify barriers and facilitators to participation in the regional STF ecosystem
- develop strategies for future-proofing the regional STF ecosystem, with particular emphasis on areas for future investment to support growth and resilience.

To achieve these goals, the research team synthesized findings from quantitative and qualitative research methods. Qualitative data informed the scope and types of quantitative analyses conducted; quantitative data analysis served both as a prompt for focus group discussion and to verify and qualify select perceptions raised. Additionally, the research team assessed the qualitative data for context to and potential explanations for patterns observed in quantitative data.

The definition of STF occupations (which we describe in detail in the next section) was drawn from a literature review, expert elicitation, and quantitative analysis of data from the U.S. Department of Labor's Occupational Information Network (O*NET) and the U.S. Bureau of Labor Statistics (BLS). Once the set of *STF occupations* was defined, we assessed the STF workforce ecosystem using current and historical data from the BLS and the U.S. Census Bureau.

We held a series of focus groups with representatives spanning three stakeholder groups, with two sessions held per group: 16 participants representing employers (founders and CEOs, human resources profes-

---

[1] We use the word *ecosystem* to describe a complex system of actors that interact within a shared space. This does not necessitate that their actions are coordinated or symbiotic.

[2] Allegheny Conference on Community Development (2017) discusses the emerging labor shortage for the Pittsburgh region, occupations and skills expected to be in demand, and regional assets.

[3] Andes et al. (2017) characterizes Pittsburgh's innovation economy as full of potential in terms of leveraging educational institutions and existing entrepreneurial supports but with specific challenges: a weak connection between research and industry, limited production of high-growth startups, and a population that is not well aligned with innovation.

[4] Despite the importance of understanding STF workforce differences across all demographic and geographic groups, we were unable to fully explore differences by race or ethnicity because of limitations in the granularity of employment data collected and the region's comparatively small number of residents of color. We also were unable to quantitatively examine several additional important dimensions of diversity, such as disability, LGBTQ+ identity, and other factors that may make up a worker's identity.

sionals, and directors of intermediary organizations), 20 participants representing education and training providers (executive directors and other senior leaders), and 16 participants representing economic development organizations (executive directors and other senior leaders). Although workers and learners were not directly represented in focus groups, representatives from community-based organizations (CBOs), community development corporations, and nonprofits were specifically selected to relay perspectives of the various populations they serve, including those of workers and learners. Focus group participants were selected to offer a diverse set of perspectives based on their personal and professional backgrounds (e.g., diversity by career category, race/ethnicity, age, gender, geography). Focus group findings represent the perspectives of participating individuals; we acknowledge that if we had included others in the focus groups, we might have heard different opinions. Throughout this report, we will describe findings from focus groups in terms of these perspectives, noting similarities and differences between stakeholder groups and participants overall when relevant.

Findings from the literature review are reported with citations to differentiate them from focus group participant perspectives. We also conducted a detailed scan of the region's postsecondary education and training programs, including certificate and degree programs at colleges and universities and short-term boot camps and apprenticeships. Finally, to complete our analysis, we drew on two proprietary data sources: job posting data from Burning Glass Technologies (now Lightcast), a private firm that provides information on labor market trends by scraping online job postings, and economic spillover data from IMPLAN. As our research progressed, two key themes emerged from the quantitative and qualitative analyses: equity (in both geography and demography) and sustainability (in terms of future-proofing the ecosystem for resilience and adaptability). These two themes serve as lenses for discussion in this report.

This report is meant to convey the high-level takeaways from our analyses in a format that is accessible to a variety of stakeholders. More-technical details on our methods and additional findings are available in the report's appendixes, which are available online (see Table 1.1).

For the purposes of this report, *the Pittsburgh region* refers to the seven counties that make up the Pittsburgh metropolitan statistical area (MSA): Allegheny (which includes the city of Pittsburgh), Armstrong, Beaver, Butler, Fayette, Washington, and Westmoreland. To contextualize Pittsburgh's present and potential future, we selected two peer regions for comparison, in consultation with our project funders. The Boston MSA was chosen as an established STF hub, and the Nashville MSA was chosen as an emerging STF region, much like Pittsburgh. Criteria for choosing Pittsburgh MSA comparators included demographics, occupational composition of employment, and local economic conditions.

Defining the Science- and Technology-Focused Workforce

There are several existing definitions of science, technology, engineering, and mathematics (STEM) occupations. Why construct a new definition of the science- and technology-focused (STF) workforce? Many existing STEM occupation definitions are delineated only by bachelor's degree field or include only occupations that require a bachelor's degree, and the term *STEM* tends to conjure up such jobs (hence our preference for *STF*).[5] Given the region's industrial mix, it was our expectation that sub-baccalaureate jobs would be of unique importance to Pittsburgh's STEM workforce. Therefore, our definition had two goals: to be relevant

---

[5] The National Science Foundation (NSF) has intentionally moved away from this approach, augmenting their STEM classification with sub-baccalaureate STEM occupations. The NSF refers to workers in these occupations as the *skilled technical workforce* and is piloting new data-collection approaches to better measure this population and its contributions to STEM. In our discussions with regional STEM experts, experts felt that the NSF and Brookings definitions were too broad.

TABLE 1.1
Appendix Tables Available Online

Appendix	Description
A	This appendix presents our methods for defining the set of occupations included in our definition of STF occupations, presents several alternative definitions, and details the American Community Survey (ACS) data analyzed for demographic comparisons of the region's residents and workers.
B	This appendix provides detail on the labor data analyzed: employment and wage statistics from the BLS and the Pennsylvania Department of Labor and Industry (DLI), employment projections by occupation from the BLS and DLI, worker data from BLS's Current Population Survey, and job posting data from Burning Glass.
C	This appendix presents our approach to and the results from our landscape scan of STF education and training providers in the region.
D	In this appendix, we describe our analysis of economic spillovers using IMPLAN data and include some additional results.
E	This appendix presents our approach to focus groups, including protocol topics, analytic approach, and findings.
F	This appendix includes a detailed review of literature on the barriers and facilitators that influence STF career choices, supplementing the perspectives shared during the focus groups.
G	This appendix provides detail on existing local and national efforts aligned with our recommendations and on regional stakeholders who are potentially positioned to lead or support expanded efforts.

to Pittsburgh in how we treated the skilled technical/sub-baccalaureate workforce and to be broad enough to encompass the variety of ways in which workers in the region contribute to the STF economy, while still being narrow enough to be meaningful.

Rather than classify occupations by job title, we took a bottom-up approach similar to that of the Brookings Institution (Rothwell, 2013), evaluating occupations by reported knowledge level and importance across domains in the O*NET.[6] After consulting with local STEM experts and reviewing the literature, we identified seven domains as core STF knowledge—biology, chemistry, physics, computers and electronics, engineering and technology, mathematics, and medicine and dentistry—and three domains as secondary STF knowledge—mechanical, production and processing, and building and construction. We classified occupations as STF occupations if they require *either* a high level of core STF knowledge *or* a moderately high level of core STF knowledge and a high level of secondary STF knowledge.[7] After adding some manual classifications to occupations that lacked knowledge data (see Appendix A), we classified 348 of 997 occupations, or nearly 35 percent, as STF occupations.[8] This is distinct from the share of total employment in STF occupations, which is 18 percent in Pittsburgh and 16 percent nationally, as of 2021.

---

[6] The O*NET surveys incumbent workers about the education and experience requirements, work context, and work activities of their jobs.

[7] At least 1.5 standard deviations above the mean in one or more core domain or at least 1.5 standard deviations above the mean in a secondary domain *and* at least 1.0 standard deviation above the mean in a core domain.

[8] In the BLS Standard Occupational Classification (SOC) context, 228 of 790 occupations (29 percent) are classified as STF.

DEFINING THE STEM WORKFORCE

Our approach is not the only way to define the STEM or STF workforce. The Brookings Institution also takes a bottom-up approach, constructing task- or knowledge-based definitions of the STEM workforce, whereas the U.S. Census Bureau definition is based on groups of related occupations. The NSF's current definition is also based on occupation group, with the addition of select sub-baccalaureate STEM occupations. When we compare the science and technology workforces under these differing definitions, the RAND STF workforce falls roughly in the middle in terms of its size, typical earnings, and key characteristics, such as workforce diversity and the share of the workforce with a bachelor's degree. Detailed comparisons can be found in Appendix A, Tables A.26 and A.27.

Table 1.2 displays the size of the STF or STEM workforce across U.S. Census Bureau, RAND, NSF, and Brookings definitions. The RAND STF definition is the second narrowest, with only the Census definition identifying a smaller workforce. More details on the demographics of the STF/STEM workforce under various definitions are available in Appendix A, Tables A.26 and A.27.

For readability, we collapsed the 348 STF occupations into occupational groupings (containing only STF occupations) for several analyses. These groupings were constructed based on the two-digit BLS SOC code. Four are neatly defined by SOC codes: computing and math; engineering; health; and life, physical, and social sciences. There are remaining STF occupations scattered across SOC codes with no obvious categorization. We divided these by educational requirements: Remaining STF occupations that generally require a bachelor's degree are categorized as "business, management, and related," and those that require sub-baccalaureate education and training are categorized as "production, maintenance, and related." A detailed description of the STF occupation groups is available in Appendix B.

TABLE 1.2

Size of the Science and Technology Workforce Using Alternative Occupational Categorization Schemes, by Metropolitan Statistical Area, 2019

Region	Pittsburgh	Boston	Nashville
U.S. Census Bureau	221,340	552,131	146,157
RAND STF workforce	**255,774**	**650,067**	**184,730**
National Science Foundation, total	332,539	770,230	262,522
Brookings "Hidden STEM," total	370,882	936,575	292,295
Total population	2,523,879	5,273,122	2,106,031

SOURCE: Features 2019 data from the ACS (U.S. Census Bureau, undated).

NOTE: *Workforce* refers to individuals ages 16 or older who are employed full or part time or who are unemployed and looking for work. More details on definitions are available in Appendix A. Regions in our ACS analyses are broader (including both more land and more people) than in our other analyses because Public Use Microdata Areas (PUMAs), the most granular level of geography identified, may combine a county in the MSA with one outside it. For example, Greene, Indiana, and Lawrence Counties are included in our ACS analyses of the Pittsburgh area.

Criteria for Assessing the STF Workforce Ecosystem

In this report, we discuss the equity and sustainability of Pittsburgh's STF workforce ecosystem. We define an *equitable regional ecosystem* as one that provides equal opportunities to prepare for, participate in, and benefit from the STF ecosystem. Such an ecosystem would address national and regional systemic barriers that have historically led to unequal participation and benefit based on such factors as race/ethnicity, gender, ability, education level, socioeconomic status, and geography (Langston, Scoggins, and Walsh, 2020; National Governors Association, 2021; Perez-Johnson and Holzer, 2021). We define a *sustainable ecosystem* as a resilient one, one that is poised to adapt and even grow amid potential technological, social, political, and economic shifts (Perez-Johnson and Holzer, 2021). In practice, an equitable and sustainable ecosystem would support a market for talent that pays sufficient wages while creating economic value for employers, make clear information about a variety of educational and career pathways accessible to potential workers, distribute the benefits from economic growth among workers and across geographic areas, and react to changing information and ecosystem needs in a regionally coordinated manner.

We examine each of these components in turn in the chapters that follow. We first assess Pittsburgh's current STF ecosystem, discussing

- the characteristics of the STF workforce
- the sufficiency of STF wages
- the sufficiency of current STF labor supply
- the availability of information to support career decisionmaking
- the climate of STF workplaces
- the educational requirements of STF jobs
- the accessibility and relevance of education and training
- the coordination of stakeholder activity and planning
- the distribution of gains from the STF economy.

We then turn to the following four future-oriented topics that could support or threaten the sustainability of Pittsburgh's STF economy:

- supporting STF employment beyond the health sector
- diversifying the STF workforce
- balancing regional needs in an era of remote work
- identifying skills for the future.

We end with a discussion of broad regional findings, their implications for future growth, and recommendations to enhance the equity and sustainability of the region's STF workforce ecosystem.

CHAPTER 2

Pittsburgh's STF Ecosystem: Recent Past and Present

In the following sections, we define and discuss the attributes of an idealized STF workforce ecosystem, which were itemized in the previous chapter. We assess Pittsburgh's current status by leveraging quantitative and qualitative data on the region's STF ecosystem. Where no such data are available, we turn to the literature to understand the factors influencing the attribute at a national level to provide insight into the Pittsburgh context. Quantitative data are generally not real-time; we discuss likely future trends where possible. Chapter 3, "Planning for the Future," exclusively focuses on prospective topics that extend analyses of current data.

Robust STF Employment

An ecosystem with robust employment would feature steady and sustainable employment growth; a diverse, skilled workforce matched with appropriate jobs; and employment across a relatively balanced set of industries to ensure resilience in times of economic downturn. This section relies on analyses conducted using data from the 2015 and 2019 ACS and the 2021 Occupational Employment and Wage Statistics (OEWS), which is available from the BLS (BLS, undated; U.S. Census Bureau, undated). We also rely on our detailed literature review of the factors that influence STF career choices.

Using the definition of STF constructed for this project, we found that there are approximately 185,640 STF workers currently employed in the Pittsburgh region (as of May 2021) (BLS, undated), accounting for 18 percent of employment in the Pittsburgh region.[1] Note that we define the *STF workforce* more broadly to include both employed and unemployed STF workers (which is reflective of potential talent in the region). From 2015 to 2019, the Pittsburgh STF workforce grew as a share of the overall labor force and became slightly more racially and ethnically diverse. The share of workers born outside Pennsylvania (including those born outside the United States) grew, as did the share holding a bachelor's degree. In this section, we compare STF workers in Pittsburgh, Nashville, and Boston; compare STF workers with non-STF workers within a region; and compare STF workers across counties within a region. It is important to note that the characteristics of STF workers may differ little from the characteristics of workers and residents in a region more generally. More-detailed comparisons between these groupings are available in Appendix A. Here, we provide a snap-

---

[1] Note that this count of *workers* differs from the *workforce* estimate presented in Table 1.2 because of the differing data source (OEWS versus ACS), recency of available data (2021 versus 2019), exclusion of unemployed or part-year workers not employed in the survey reference month (who reflect potential talent in the region), and other methodological reasons. For example, our ACS analyses include three counties outside the Pittsburgh MSA because three PUMAs (the most granular level of geography available) include both an in-MSA and an out-of-MSA county. Additionally, because ACS occupational categories tend to be less granular than the categories used in the OEWS, in some instances, an occupation that is not STF in the BLS OEWS analyses is included in an ACS occupational group that is determined to be a STF group. See Appendix A for more details.

shot of who the STF workers in the Pittsburgh region are and how their characteristics compare with those of workers in peer regions.

In 2019, about 60 percent of Pittsburgh's STF workforce (ages 25 and older) held at least a bachelor's degree. This rate is comparable to Nashville's (59 percent), but lower than Boston's share (74 percent), which has a more–highly educated workforce in general (more than half of Boston workers—STF and non-STF combined—hold at least a bachelor's degree, compared with 43 percent of workers in Pittsburgh).[2]

Like Pittsburgh's Overall Population, the Region's STF Workforce Is Older and Less Racially and Ethnically Diverse Than in Peer Regions

Eighty-six percent of the region's STF workers are White and non-Hispanic, which is a larger share than in Boston (72 percent) or Nashville (76 percent) (see Figure 2.1). The region has a particularly low share of Hispanic workers (about 2 percent of the STF workforce). Like in peer regions, 44 percent of Pittsburgh's STF workers are women. Pittsburgh's STF workforce is also older than in the peer regions, with an average age of 43. Relative to households with only non-STF workers, households with STF workers (across Pittsburgh, Boston, and Nashville) were more likely to be married-couple households, to have children under age 18, and to be living in an owned home (as opposed to renting).

Looking across the Pittsburgh region, STF workers in Allegheny County are more likely to be workers of color, be young, be born out of state, and hold a bachelor's degree than their STF counterparts in the surrounding counties. About half of STF workers in the surrounding counties (ages 25 and older) do not hold a bachelor's degree (and two-thirds of STF and non-STF workers in the labor force). Households with STF workers in the surrounding counties are even more likely than their Allegheny County counterparts to be married, have children under age 18, and be homeowners.

Health Dominates Pittsburgh's STF Employment

As shown in Table 2.1, Pittsburgh's health sector boasts an outsized share of STF employment—43 percent of STF workers (and 7.7 percent of all workers) work in health STF occupations. Health is a similarly large share of STF employment in Nashville (45 percent), but health constitutes a smaller STF share in Boston (33 percent) and in the United States overall (38 percent). Computing and mathematical occupations are the second-largest grouping of STF employment in the three focal regions and in the United States overall. However, in this category, Pittsburgh (18 percent of STF employment) lags Boston (24 percent) while slightly lagging the United States overall (20 percent). Pittsburgh has a comparatively larger share employed in architecture and engineering occupations (12 percent of STF workers), on par with Boston and the U.S. average, while exceeding Nashville's share of 8 percent.

The share of Pittsburgh STF workers in computing and math occupations has increased in recent years, while the share in health occupations and production and related occupations has declined. Although there is no clear optimal distribution of STF workers across occupational groups, supporting growth in non-health STF occupations may be advantageous; we found larger economic spillovers and more-competitive markets for workers outside health.

---

[2] The workforce consists of those ages 25 to 64. When we narrow this to people aged 25 to 44, Pittsburgh still lags Boston in share of the population with a bachelor's degree but moves further ahead of Nashville, suggesting that age composition plays a role.

FIGURE 2.1

Demographics of STF Workers

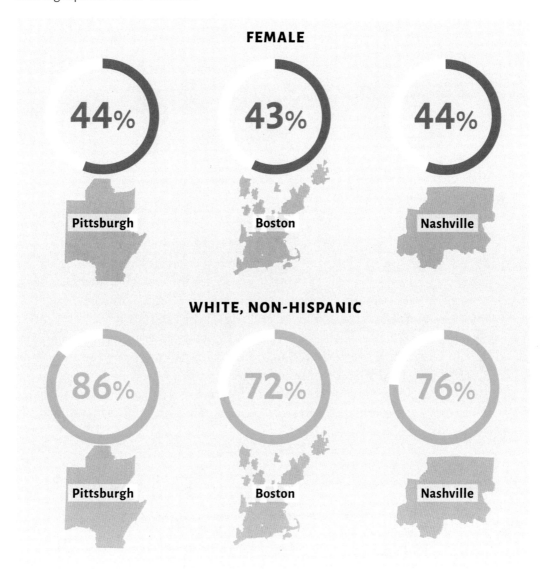

SOURCE: Features 2019 data from the ACS (U.S. Census Bureau, undated).

TABLE 2.1

Composition of STF Employment, by Occupational Group

	All STF	Business, etc.	Computing and Math	Architecture and Engineering	Health	Production, etc.	Sciences
Boston	21%	15%	24%	11%	33%	9%	8%
Nashville	16%	13%	17%	8%	45%	14%	3%
Pittsburgh	18%	11%	18%	12%	43%	12%	4%
United States	16%	13%	20%	11%	38%	14%	4%

SOURCE: Features 2021 data from the OEWS (BLS, undated).
NOTE: The first column represents the share of STF employment relative to all employment. All other columns represent the occupation group's share of STF employment. See the section titled "Defining the Science- and Technology-Focused Workforce" in Chapter 1 for occupational group definitions.

Sufficient Wages

Members of the focus groups reported, and prior research corroborates (e.g., Wiswall and Zafar, 2015), that perceived and actual wages for STEM jobs must be sufficiently competitive to incentivize individuals to pursue the required education and to select specific jobs. In this section, we assess wages through two lenses: (1) Are these wages sufficient relative to the local cost of living? and (2) Are they competitive enough to attract a national labor pool? If the Pittsburgh STF ecosystem continues to grow and the Pittsburgh region's population continues to age and shrink, the region will need to engage talent from outside Pittsburgh.

We define key terms related to our discussion of wages in the box.

Of the regions in our analysis, Boston has the highest cost of living (53 percent above the index average),[3] Nashville has the lowest (6.5 percent below the index average), and Pittsburgh is in the middle (3.4 percent

---

KEY TERMS RELATED TO WAGES

A *family-sustaining wage* sufficiently covers necessary expenses, such as food, child care, housing, transportation, utilities, and medical expenses, without needing supplemental support (e.g., drawing down savings, receiving public transfers) for a family of a given size (in this report, we consider a family of four with two earners).

This is distinct from a *living wage*, which focuses on these expenses for a single person. We recognize that family structures vary and that not all workers intend to have children. However, given the low cost of housing in Pittsburgh and the comparatively high cost of other per-person necessities, our definition of a *family-sustaining wage* is a more conservative measure of sufficient wages.

We also discuss *cost-of-living–adjusted wages*, scaling wages by a region's average cost of living (i.e., necessary expenses). Pittsburgh's cost of living is 103.4 percent of the index average, meaning that to match the buying power of $100 nationally, Pittsburgh workers need to make $103.40.

---

[3]   The national average reflects the average of areas participating in the Council for Community and Economic Research's Cost of Living Index. Participation is open to all MSAs and to non-MSA areas with county populations above 50,000 and city populations above 35,000. The index excludes sales, income, property, and other taxes.

above the index average). Pittsburgh's above-average costs for groceries, utilities, and transportation offset the region's lower cost of housing.[4]

Pittsburgh STF Workers Make Less Than Average, Even After Adjusting for Cost of Living

According to our analysis of 2021 OEWS data, STF workers in Pittsburgh make an average of 8 percent less than the national average for their occupation (see Figure 2.2). Nashville's STF workers fare worse—making 11 percent less than the national average—but Boston workers take in 13 percent *more* than the national average. This "Pittsburgh discount" is less pronounced for Pittsburgh's non-STF workers: Although they have lower average salaries than STF workers ($50,350 versus $69,120), they make only 4 percent less than their national counterparts. All three regions have cost-of-living–adjusted wages that are below the national average, with Nashville having the highest.[5] Wages adjusted for cost of living account for regional differences in common expenses, such as home prices, groceries, and child care.

Even with this "discount," most Pittsburgh STF jobs pay family-sustaining wages locally. Ninety-five percent of STF workers work in an occupation where the average salary makes the median-priced home affordable for an individual earner (i.e., where the monthly mortgage payment is less than one-fourth of income), compared with 63 percent of STF workers in the United States and just 30 percent in Boston. Indeed, Pittsburgh has been noted globally for the relative affordability of its housing (Cox and Lucyshyn, 2022), and nearly half of Pittsburgh non-STF workers can also afford that monthly mortgage payment.

There is some variation in pay competitiveness by occupation group. The Pittsburgh discount ranges from a low of 1 percent in production and related sub-baccalaureate STF jobs to a high of 11 percent in computing and mathematics occupations. Pittsburgh's health workers are also paid about 10 percent less than

FIGURE 2.2

Pittsburgh's Sufficient—but Noncompetitive—STF Wages

8% LESS

Pittsburgh STF workers are paid **8 percent less** than the national average.

95%

95 percent of Pittsburgh STF workers can afford a house.

94%

94 percent of Pittsburgh STF employment is in occupations that tend to pay a living wage.

SOURCE: Features 2021 data from the OEWS (BLS, undated).

---

[4] This is a point-in-time, cross-region comparison: Costs of these goods and services have increased throughout the United States during the pandemic; however, this refers to Pittsburgh's cost *level* being above that of other regions (in addition to the upward trend everywhere).

[5] It is unknown how strongly workers consider local cost of living when comparing job opportunities. Some expenses are fixed regardless of job location (e.g., student loan payments), and areas with a high cost of living tend to have attractive amenities that may offset those costs in the view of the worker.

their national counterparts (in Boston, health workers earn a 15-percent premium). If we exclude health occupations—the largest grouping—from our STF wage calculation, the Pittsburgh discount shrinks to 6 percent (whereas Nashville's discount expands to 13 percent).

Most STF Workers Across the Three Regions Can Afford Basic Family Needs

Although Pittsburgh STF workers make less than the national average, the average Pittsburgh STF worker in a two-income household is more likely to be able to support a family of four than Pittsburgh's non-STF workers. Looking at this from a household perspective and incorporating a broader variety of costs, such as taxes, child care, and food, 94 percent of Pittsburgh STF workers earn a salary such that two earners at that salary could support a family of four.[6] This is consistently true for STF workers regardless of community— Pittsburgh's rate is similar to the share in Boston (92 percent) and in Nashville (90 percent), but only 47 percent of Pittsburgh's non-STF workers meet this threshold.

Wages are somewhat consistent across Pittsburgh MSA counties. For the three STF occupations for which all seven counties had sufficient employment to report wages, the lowest and highest counties' wages were within 20 percent of the MSA average (see Appendix B for details). Looking at occupational groups, we see more variation, but this appears to be driven more by composition of the occupation group (e.g., whether the county has a hospital employing many surgeons and specialists or if the health care practitioners in the county are physicians and physician assistants) than by variation in wages outright.

In sum, Pittsburgh STF jobs pay enough to buy a house as a single worker and, much like their national counterparts, these jobs pay enough to support a family in a dual-income household. Yet Pittsburgh STF wages are not competitive when compared with the wages of similar jobs in other regions, particularly in such growth-targeted areas as computing and math occupations; in the region's comparatively smaller occupation groups, such as business and related occupations; and in highly employer-concentrated occupations, such as health. If Pittsburgh's market for STF labor becomes more exposed to national influence (e.g., through an increase in the availability of remote work or additional sourcing of STF talent from outside the region), wages will likely need to increase for Pittsburgh employers to remain competitive hirers.[7]

Sufficient Labor Supply

A sufficient labor supply is one that has enough appropriately skilled individuals to sustain future growth. Pittsburgh struggles with this in several dimensions—the region's population has shrunk in recent decades (in addition to the sharp decline in population following the decline of regional manufacturing). The region is aging and lacks younger workers and families, struggles to supply and retain mid-career technical and managerial talent (as shown in Figure 2.3, reported by focus groups, and relayed in Tripp et al., 2021), and lacks the newcomers and demographic diversity that are often engines of innovation. In this section, we combine analyses of Census Bureau Population Estimates (2010 and 2019) with data from the ACS (2015 and

---

[6] Although Pittsburgh is very cost-competitive on housing, other essential goods and services are comparatively more expensive, such that overall cost of living is slightly higher than in Nashville. For example, the Massachusetts Institute of Technology's living wage threshold for Pittsburgh is $43,120, compared with $40,100 in Nashville and $60,300 in Boston (Massachusetts Institute of Technology, undated).

[7] It is important to consider the additional ramifications of higher STF wages in the Pittsburgh region. For instance, higher wages could drive up housing prices. An influx of remote STF workers moving to Pittsburgh for the low housing costs would drive up prices as well. Higher housing costs may make the region less appealing, causing some current residents to move and slowing the flow of in-migration from other regions. On the other hand, current homeowners may see equity grow as home prices rise, creating more wealth for current STF and non-STF workers.

FIGURE 2.3

Population Change, by Age Group, 2010–2019

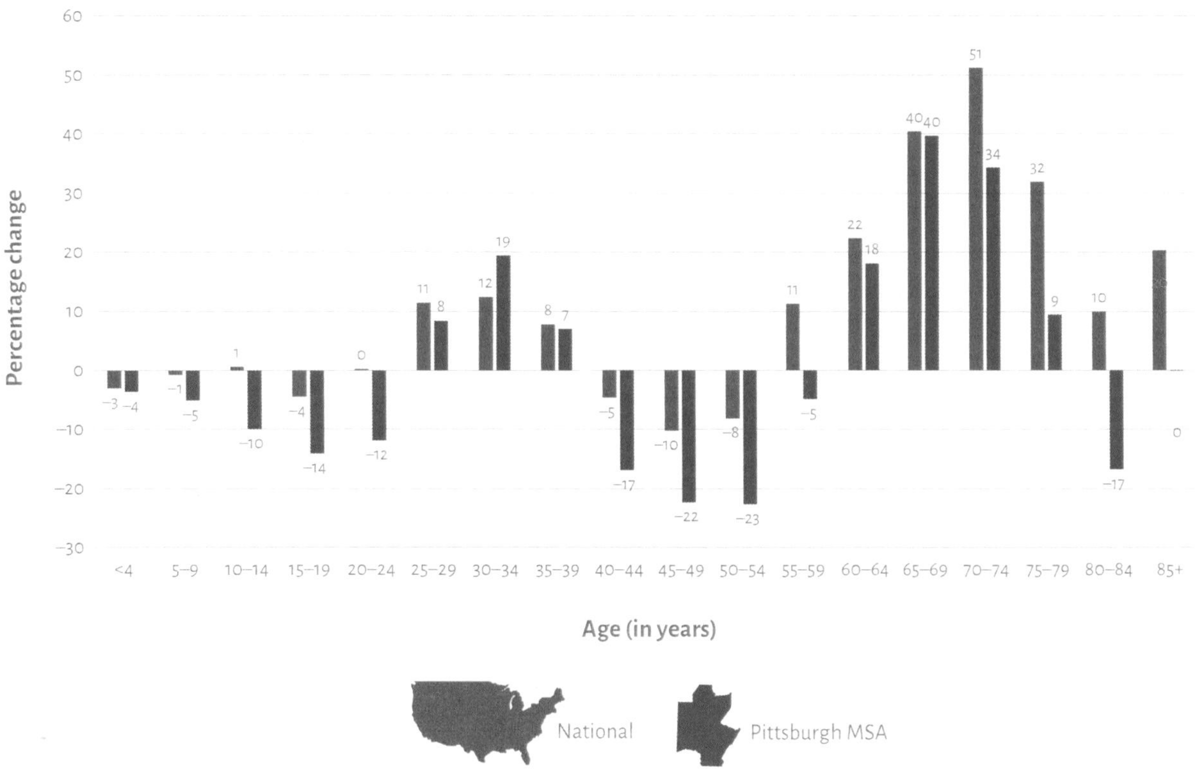

SOURCE: Features 2010 and 2019 data from the ACS (U.S. Census Bureau, undated).

2019), the Current Population Survey (2021), Burning Glass, and U.S. Citizenship and Immigration Services, along with findings from focus groups and the literature review to assess the sufficiency of the region's labor supply.

Pittsburgh's Population is Shrinking and Aging

The Pittsburgh region (as of 2019) had roughly the same share of its population in the prime worker age group (ages 25 to 54) as the United States overall: 38 percent in Pittsburgh versus 39 percent nationally. But outside that core group, Pittsburgh lacks the young workers and future workers necessary to sustain a workforce: Thirty-five percent of Pittsburgh's population is aged 55 or older, which is notably higher than the 29 percent nationally. Conversely, just 27 percent of the region's population is younger than 25, versus 32 percent nationally. Figure 2.3 depicts the change in population in the region by age group in comparison with national trends. Over the past decade, the Pittsburgh region lost residents under the age of 25, shrinking those age groups at a much faster rate than in the United States overall (which saw very slight declines). The region's number of people ages 25 to 39 grew overall, and within the 30–34 age group, the region exceeded the national growth rate. However, this growth was surpassed by substantial declines in the 40–54 age group that were more than twice the rate of national decline. Pittsburgh's population of residents ages 60 to 79 is also growing, which is similar to the United States as a whole as the large baby boomer cohort ages.

Overall, Pittsburgh does not struggle with labor force participation more than other regions or the United States at large: The region lags the national participation rate by about 1 percentage point. However, two populations do stand out locally—older residents' labor force participation rate and rates of mothers of children under age 18 working full-time, year-round.[8] Pittsburgh residents between the ages of 55 and 75 have a lower labor force participation rate (43 percent) than their counterparts in Boston, Nashville, or the United States overall. This is driven primarily by a higher rate of retirement and a slightly older average age within this age band. Given the disproportionate size and growth rate of this group in Pittsburgh, encouraging potential retirees to delay retirement could temporarily boost labor supply.

Full-time, year-round employment rates among mothers of children, especially young children, are lower in Pittsburgh than in peer regions. Although about three-quarters of Pittsburgh mothers aged 25 to 39 were in the labor force, just 47 percent worked full-time, year-round, compared with 50 percent in Boston and 54 percent in Nashville. This rate is 3 percentage points lower when we restrict the sample to mothers of children younger than six, with just 44 percent of Pittsburgh mothers in this group working full-time, year-round. Lack of access to affordable, high-quality child care came up repeatedly in focus group conversations across stakeholder groups as a barrier to both working and pursuing further education and training.[9] (Detailed findings from focus groups, organized by theme, are in Appendix E.) Child care access also emerged in the literature review as a barrier or facilitator (depending on the level of access to child care) to participation in STEM (see Appendix F for detailed findings). Perceived or true inaccessibility of child care may also drive mid-career workers out of the Pittsburgh region and discourage relocation to Pittsburgh; this is especially problematic because focus group employers cited mid-career workers as a very challenging group to attract, develop, and retain.

Pittsburgh Lacks the Population Inflows That Can Contribute to Innovation

Boasting universities with international reputations, Pittsburgh has the potential to draw workers from around the globe. However, Pittsburgh's STF labor force has a smaller share of foreign-born workers—and workers born out of state—than the STF workforces in Boston (in particular) and Nashville. Immigrants and transplants, who often bring different knowledge, capabilities, and cultural context, can be important drivers of innovation (Burchardi et al., 2020; Hunt and Gauthier-Liselle, 2010; Niebuhr, 2010). Pittsburgh-area employers have less than half the per capita rate of H-1B visa approvals as do Boston employers, with just 5.8 approvals per 1,000 residents compared with 16.4 approvals per 1,000 in Boston. (Nashville slightly lags Pittsburgh with 4.9 approvals per 1,000 residents.)[10] Pittsburgh's STF workforce, though, is notably more likely to be foreign-born than the Pittsburgh labor force overall, which is not the case in the comparator cities (their STF workers are about as likely to be foreign-born as their labor forces overall). This suggests that Pittsburgh STF employers may use hiring strategies that are different from those of their non-STF counterparts.

Data from Burning Glass shed light on which occupations may be experiencing the most-pronounced worker shortages. Specifically, the average number of days it takes to fill job openings and the platform's

---

[8] We present data drawn from two separate sources in this section. Labor force participation rates (the share of individuals either employed or unemployed and actively seeking work) for older workers are drawn from the Current Population Survey Merged Outgoing Rotational Groups files and reflect 2021 rates (see Appendix B, Table B.30). This sample is too small to analyze mothers of young children. For this analysis, we use ACS data from 2019, focusing on rates of full-time, year-round employment (see Appendix A, Table A.11). Both analyses reflect the full population in the specified age and presence of children groupings. Although understanding variation in labor force and full-time employment rates by factors such as race/ethnicity, educational attainment, or income level could shed light on where challenges are more pronounced (or suggest explanations for observed disparities), the sample sizes for the Pittsburgh MSA preclude these more-granular analyses.

[9] It is also possible that Pittsburgh's lower cost of living makes single-income family households more feasible.

[10] H-1B data for the past five federal government fiscal years (through September 2021) were drawn from U.S. Citizenship and Immigration Services (undated). Data are reported at the zip code level and were crosswalked to MSA boundaries. Adjustment to per 1,000 residents used 2019 ACS population data for the MSA.

categorizations of occupations as "easier" or "harder" to fill relative to national patterns are two metrics that can provide insight into potential worker shortages. Occupations that were harder or much harder to fill in the Pittsburgh area from 2019 to early 2022 (and that had high or medium demand based on the number of job listings) include a couple of managerial occupations (construction managers and natural sciences managers); health occupations, such as surgeons and medical records and health information technicians; and computer-controlled machine tool operators (metal and plastic). See the box for a list of such hard-to-fill jobs. Appendix B (and specifically Table B.35) includes more-detailed information on hard-to-fill occupations.

As local companies grow, there will be a need for more talent, younger talent, and more-diverse talent, all with some STF knowledge. If the region is unable to attract, train, and retain these workers, worker shortages may limit the growth of companies with a significant presence in Pittsburgh and, in turn, further limit the growth of the STF workforce in Pittsburgh in a perpetuating cycle.

Available Information to Inform Potential STF Workers' Career Decisions

Information about STF careers and career pathways, and targeted advertising and recruitment efforts by postsecondary institutions, can increase the likelihood of individuals choosing an STF career (see Appendixes E and F for more detail). However, high school students and other potential STF workers in the region might lack the information about or hold assumptions or beliefs that prevent them from seeking employment in STF careers. To assess how well-informed potential decisionmakers in the Pittsburgh region are, we drew on our focus group data, literature review, and scan of the region's STF education and training programs. For example, in our focus groups, stakeholders believed that negative perceptions and misconceptions about what constitutes a "tech job" on the part of potential workers (including high school students) can be limiting factors in recruitment (by both educational programs and employers) and hiring. Focus group participants believed that potential workers were not always aware of the variety of STF employers (e.g., that

---

HARD-TO-FILL JOBS

STF jobs that are (1) in relatively high demand in the Pittsburgh region and (2) harder or much harder to fill locally than nationally and in comparison regions include

- veterinary technologist and technicians
- family and general practitioners
- computer-controlled machine tool operators (metal and plastic)
- search marketing strategists
- construction managers
- medical scientists, except epidemiologists
- hospitalists
- obstetricians and gynecologists
- surgeons
- cardiovascular technologists and technicians
- medical records and health information technologists.

STF jobs are also found among such employers as supermarket Giant Eagle and law firms), the education and skills needed for entry into STF careers, and STF career pathways within employers or sectors.[11]

Potential STF Workers Can Be Deterred by Perceptions of What STF Jobs Require and Who They Are For

Several focus group participants representing education and training providers described direct experiences with potential STF workers expressing concerns that STF jobs require many years of education and very specialized skills (although our analysis indicates that a sizable share of the region's STF jobs do not have these requirements). A few participants expressed concern that this belief may contribute to potential workers' lack of interest and confidence in developing STF skills, with the result that potential workers may prefer non-STF jobs paying good wages and with more familiar—or accessible—qualifications and career pathways. For example, a focus group participant from an economic development organization said that the relatively high starting wages paid to clerks at a regional convenience store chain is an example of a visible, well-understood job with immediate benefits that some workers might find more attractive than embarking on an uncertain path to an STF career with uncertain future benefits.

Some focus group participants also perceived that youth of color and those raised in low-income families often lack exposure to STF jobs and reported seeing few people like them in those positions. Together, these experiences contribute to the perception that, in the words of one focus group member, "[STF] jobs aren't for me." Research on the national STEM labor market suggests that stereotype threat, which can lead individuals to behave in a way that conforms to widely held stereotypes about their race, ethnicity, or gender, may also prevent members of underrepresented groups from engaging in STF education and training (Ong, Jaumot-Pascual, and Ko, 2020; Totonchi et al., 2021). Nationally, these factors have been shown to contribute to developing an identity that excludes STF careers (Mahadeo, Hazari, and Potvin, 2020; McLean, Koepf, and Lilgendahl, 2022).

Specific to the Pittsburgh region, the legacy of the industrial economy may impede recruiting workers into current manufacturing jobs. Employers and education and training providers who participated in focus groups reported believing that potential workers perceived manufacturing jobs, in the words of one participant, to be "dirty jobs" based on the historical role of the steel industry in the regional manufacturing economy. In truth, focus group participants told us that technological innovation, including automation, has rapidly changed the skills required to work in advanced manufacturing, and participants across all three stakeholder groups said that potential workers and education and training providers may not be aware of the STF skills required for these jobs.

Overcoming informational barriers can be complex and may require multiple approaches. Career counseling in high schools (Mau and Li, 2018; Shoffner et al., 2015; Tsui, 2007) and targeted recruitment efforts by undergraduate programs and employers (Beninson and Alper, 2021; Deng, Zaza, and Armstrong, 2020; Holloman et al., 2021; Tsui, 2007; White and Shakibnia, 2019) have been found to increase student interest and participation in STF programs. Prior literature has also established that those who anticipate positive outcomes in STF careers (such as success, enjoyment, and financial stability) are more likely to pursue STF careers (Morales-Chicas et al., 2021; Ong, Jaumot-Pascual, and Ko, 2020; Shoffner et al., 2015). This suggests

---

[11] These perceptions were shared by education and training providers (based on direct experiences working with learners and prospective workers) and by employers and economic development organizations (based on assumption or inference).

the need to clarify STF career pathways and their associated benefits, particularly for members of under-represented groups.[12]

Pennsylvania Lags Peers in Tracking Outcomes for Program Completers

The Pittsburgh region lacks a way to reliably track completion and employment for learners who do engage with local education and training programs. The limited, inconsistent, and siloed nature of available data makes it difficult not only to assess the effectiveness of local education and training programs but also to describe actual STF career pathways for students or demonstrate the real-world value of training programs to potential learners.

Pennsylvania's statewide longitudinal data system—known as the Pennsylvania Information Management System—collects limited data on postsecondary education, does not connect to workforce data, and is difficult to access for both researchers and practitioners.[13] Moreover, this system is separate and siloed from data collected on outcomes for participants in programs on Pennsylvania's Eligible Training Provider List for the Workforce Innovation and Opportunity Act.[14] Pennsylvania lags Kentucky, West Virginia, Texas, and Minnesota in tracking education and training program outcomes, linking these data with kindergarten through grade 12 (K–12) and other data systems, and providing access to practitioners and researchers. And although national Department of Education data included state of residence at the point of admission

> "We capture data when students are active in our program . . . what their career plans are at that time. But when they graduate high school, for us to follow them afterwards . . . we have not found a successful way to do that. And speaking with colleagues that are in similar programs in that K to 12 space . . . you know, four or five years down the line, where are those students? We just haven't figured out how to do that yet. If someone can solve that, that would be amazing."
> —Education and training focus group participant

---

[12] For our analyses, we defined *underrepresented groups* as those groups participating in the STF ecosystem at disproportionately low rates. In the Pittsburgh region, this includes Black, Hispanic/Latino, and female residents, in addition to residents of the surrounding counties (see Figure 2.1 and Appendix A for supporting statistics).

[13] According to the Pennsylvania Department of Education, currently, the department collects only "unit-level data that are mandated by federal statute or regulation" (Pennsylvania Department of Education, undated). This includes, for example, data collections on students participating in postsecondary career and technical education programs under the Strengthening Career and Technical Education for the 21st Century Act of 2018.

[14] Outcomes for participants in programs on the Eligible Training Provider List are reported in the aggregate (i.e., they are not linked to other characteristics of participants or their prior education or training records) and include the "performance measures" of credential attainment, employment rates two and four quarters after exiting the program, and median earnings (Pennsylvania DLI, 2020).

of *incoming* undergraduate students, there is not a comparable dataset that tracks where students go upon graduation, either geographically or professionally.[15] According to focus group participants, individual education and training program providers struggle to keep track of where participants move or are employed when they leave or graduate. Although colleges and universities may track some of this information for graduates, learner outcomes at many other programs are not tracked systematically (if at all) and generally are not standardized and linked across institutions. To credibly address the misinformation about STF careers discussed previously and present learners with realistic potential outcomes, local and/or state postsecondary education and training and workforce data collection efforts would need to improve.

Welcoming and Talent-Nurturing Workplaces

Organizational culture affects worker recruitment, worker success, and worker retention, all of which shape the STF workforce ecosystem and its productivity (Cech and Pham, 2017; Makarem and Wang, 2020; McGee and Martin, 2011; Muñoz and Villanueva, 2022). Organizational culture is of particular interest for some STF sectors: Prior research has found that, nationally, tech and manufacturing sectors have a reputation for being unwelcoming to workers from underrepresented groups, such as women, workers of color, and those from lower socioeconomic status (e.g., Benner, 2017; Chokshi, 2022; Marín-Spiotta et al., 2020; Yonemura and Wilson, 2016). Because there are few systematic quantitative datasets measuring workplace culture, this section relies on findings from our focus groups and literature review.[16] Appendixes E and F provide more discussion of the importance of organizational culture in attracting and retaining workers.

Given that demand for STF talent in Pittsburgh likely will outpace what the region produces and retains, Pittsburgh-based employers probably will engage with the national pipeline for STF workers. In this section, we draw on prior literature that discusses some of the challenges with drawing potential workers into STF training, careers, and workplaces across the United States and contextualizes this national evidence with focus group perceptions of local challenges. We first describe challenges faced by STF employers universally and then move to the potential impact of employer choices.

Positioning STF Workplaces as Welcoming Is a National Challenge

Our literature review found that nationally, STEM employers are not perceived by members of underrepresented groups (such as workers of color and women) as welcoming places where they can be successful (e.g., Benner, 2017; Chokshi, 2022; Marín-Spiotta et al., 2020; Yonemura and Wilson, 2016). Recent media reports have highlighted accounts of sexism and racism in prominent technology companies in other regions (e.g., Kerpen, 2021; Wakabayashi, 2022). Exclusionary or biased policies and procedures (particularly policies and procedures related to recruitment, minimum standards in job descriptions, and promotion) have been found to negatively affect members from underrepresented groups in STF postsecondary education and career settings (Annabi and Lebovitz, 2018; Holloman et al., 2021; Makarem and Wang, 2020; McGee and Martin, 2011). For example, the literature indicates that organizational cultures that reinforce gender stereotypes (e.g., women being encouraged to take notes in meetings or organize social events) can inhibit women's STEM careers by focusing their attention on nonpromotable tasks or discouraging them from STEM

---

[15] One exception is a pilot effort, the Post-Secondary Employment Outcomes project, that details employment outcomes for the graduates of 200 public four-year institutions. The data do not cover other types of institutions and are limited to 17 U.S. states. Details can be found in Conzelmann et al., 2022.

[16] Industry- or occupation-specific turnover is not measurable at the regional level, so we were unable to quantitatively corroborate focus group perceptions of a higher rate of STF worker turnover locally.

entirely (Makarem and Wang, 2020). Our focus group participants also perceived that at least some people of color and low-income individuals in the region do not think of Pittsburgh's technology and manufacturing sectors as welcoming to them.

These challenges extend to such STF fields as health and computing and math occupations, which are some of the most-prominent STF occupations in Pittsburgh (as shown in Table 2.1 in Chapter 2). Nationally, unexplained wage gaps persist between health care workers of different races in the United States, separate from choice of occupation (Frogner and Schwartz, 2021). These unexplained wage gaps suggest that otherwise similar workers may face different obstacles in the health sector. Additionally, the health care jobs that pay very well, such as practitioner jobs, tend to be dominated by men, in contrast to health care jobs overall (Day and Christnacht, 2019). Computing is a gender-imbalanced field as well, and evidence from the literature suggests that perceptions of the experience of women and workers of color in the field may create further barriers. A Pew Research Center survey revealed that 74 percent of surveyed women in computer jobs felt that they had experienced gender-based discrimination, compared with 16 percent of men in those jobs (Funk and Parker, 2018). Our focus group participants shared perspectives that are consistent with these national findings. Regardless of the on-the-ground reality in Pittsburgh, the region will need to contend with these national perceptions as it pursues workers from a national talent pool.

Organizational Policy Choices Influence Worker Perceptions and Worker Choices

Despite significant efforts on the part of regional employers to recruit and retain diverse candidates, most regional stakeholders we spoke to predicted an uphill battle because of a lack of demographic diversity in the region, the perceived difficulty of attracting professionals of color to the region, and perceived barriers to members of underrepresented groups accessing and persisting in STF education and training opportunities.

Employers set policies and institute procedures that contribute to their organizational climate and culture. Organizational culture came up in focus group discussions across all three stakeholder groups as a dimension of the barriers to STF employment and advancement, and participant comments focused on biases against people from underrepresented groups. For example, participants reported that people from underrepresented groups (including women) perceive that they are not offered the same opportunities within their firms as White men and often fail to see themselves reflected in higher-level positions. One participant recounted confronting their own implicit biases, realizing that they gravitated toward candidates from a similar background. Even in the absence of overt discrimination, participants perceived that workers from underrepresented groups often struggle to find a sense of community or belonging in some STF workplaces.

Organizations seeking to improve their internal culture and climate and concomitantly diversify their workforces have found some success by revising and/or adopting practices and policies to foster diversity, equity, and inclusion (DEI). Examples of such practices and policies from the literature include interventions for staff and management (e.g., implicit bias training and one-on-one mentoring)[17] and changes in hiring and management policies (e.g., skills-based hiring, competency-based job descriptions, and performance management reforms)[18] (Makarem and Wang, 2020).

---

[17] *Implicit bias* refers to when people act on prejudicial attitudes or stereotypes without conscious intent. Implicit bias training attempts to make people aware of their prejudice and provides tools to adjust their behavior.

[18] Performance management approaches affect the tone and culture of a workplace. For example, if expectations are unclear, evaluations are conducted by a single reviewer, or prejudicial language goes unchecked, people may perceive or experience bias.

Another option for improving culture and climate is clearly stating the pathways for advancement. The perceptions of local focus group participants suggest that pathways for promotion and advancement are not always clear or supported in the tech industry, which is a segment of the STF ecosystem that is increasing accessibility through noncollege pathways but has had difficulty retaining workers (as documented nationally and reported in our focus groups locally). When asked about how they communicate with workers about advancement opportunities, a small subset of employer focus group participants described efforts to clarify job descriptions and internal advancement opportunities, but a larger number of focus group participants reported that no such practices existed at their firms. If workers do not see a clear career trajectory with their current employer, they may seek opportunities elsewhere. Indeed, local employers (including three representatives of local tech firms) described a culture of high regional turnover, where workers jump from one company to another in search of advancement opportunities.

These perceptions are consistent with findings in the literature on turnover among high-tech firms in Silicon Valley (Baron, Hannan, and Burton, 2001) and with reporting in the popular press, which reports that worker tenure is low even at major tech firms, such as Amazon and Google (Johnson, 2018). Frequent turnover has been part of the national technology labor market in the sector for decades, and our focus group participants perceived that turnover is more pronounced in the Pittsburgh region. Focus group participants indicated that local tech firms fear that they act as training grounds for workers to learn and develop the necessary skills to ultimately join big tech firms, many of which are located in other markets, that offer better pay and benefits.

Investment in workers increases employee engagement (McManus and Mosca, 2015) and decreases worker turnover (Manchester, 2012). But this is a catch-22 of sorts: A subset of focus group participants indicated that high rates of turnover inhibit investments in internal talent development, particularly at startups and small companies, out of fear that the employer will not recoup the investment by the time the worker leaves the company.[19]

Organizational practices that expand the hiring pool to individuals without degrees could be particularly impactful in Pittsburgh, given the region's comparatively lower rate of bachelor's degrees (relative to Boston) and the comparatively high share of regional STF jobs with tasks that do not appear to necessitate a bachelor's degree. *Skills-based hiring* is the practice of screening and hiring new employees based on the

> "On one end, they're considering, 'How do I grow my talent?' And on the other side is . . . being a little risk-averse about growing your talent, and then getting that talent picked off by somebody else."
> —Employer focus group participant

---

[19] The inhibited investment includes the training required to advance within the company, not just training of value to competing employers. High turnover decreases the likelihood that the employer recoups the costs of investment, even if the investment itself does not make the worker more attractive to competitors.

skills needed to successfully perform the job rather than on educational background or degree. Skills-based hiring typically results in a larger pool of eligible candidates. It does not guarantee more diversity, but it can reduce barriers to entry and thus contribute to a more diverse organization. In the next section, we discuss skills-based hiring in Pittsburgh.

Like skills-based hiring, competency-based job descriptions emphasize the knowledge and skills needed to be successful in a position rather than focus on the elements of the job itself. For example, a traditional job description might list a worker's tasks, whereas a competency-based job description would describe the skills or experience the worker would need to have or obtain to be successful in the position. Competency-based job descriptions can also be used to help managers evaluate internal job candidates and to provide information to employees about the skills they need to advance or change positions, potentially countervailing previous differences of opportunity.

Appropriate Educational Requirements

Postsecondary education requirements can align worker preparation with job requirements but are also a barrier to potential STF workers. To examine the educational requirements of Pittsburgh's STF jobs, we used data on job posting requirements from Burning Glass,[20] workers' current educational attainment via the ACS, and the educational requirements of current regional employment and growing jobs as assessed by the BLS (i.e., O*NET employment projections [BLS, undated]).

Sub-Baccalaureate STF Jobs and Workers Are a Notable Part of Pittsburgh's Present and Future

Sub-baccalaureate jobs constitute a nontrivial part of Pittsburgh's employment in both STF and non-STF occupations (Figure 2.4). Thirty percent of Pittsburgh's STF employment is in jobs that do not require a bachelor's degree (per O*NET), and five of the top ten STF occupations projected to have the most openings do not require a bachelor's degree. Relative to Pittsburgh, Boston's STF employment is more concentrated among jobs requiring a bachelor's degree, which is aligned with the greater rate of degree-holding in that region. Where educational requirements are included with hiring ads, Pittsburgh postings are less likely to require a bachelor's for STF employment than employers in Boston are (two-thirds of Pittsburgh STF job postings versus three-quarters of Boston STF job postings). This may reflect differing occupational composition between regions rather than differing employer credential standards.

But there is substantial variation in degree-holding within Pittsburgh. About half (51 percent) of STF workers in the surrounding counties do not have a bachelor's degree, compared with 32 percent of Allegheny County STF workers.[21] According to several education and training provider focus group participants, many workers who hold or are pursuing a sub-baccalaureate credential may mistakenly assume stricter credential requirements (such as a bachelor's degree) for STF jobs, which may limit their pursuit of these jobs. In part, these perceptions may be reinforced by hiring norms in some STF occupations. For instance, a representative of an economic development organization noted that, despite the proliferation of local boot camps and

---

[20] The number of job postings for an occupation in an area does not necessarily correspond with the number of job openings in that occupation in the area (e.g., one posting may represent multiple jobs).

[21] It is unclear from our data whether this differing rate reflects geographic sorting after completing education or if educational attainment intentions are related to geography. Prior literature supports both explanations.

FIGURE 2.4

FIGURE 2.4
Sub-Baccalaureate STF Occupations Are Key Contributors to Pittsburgh Regional STF Employment and Job Growth

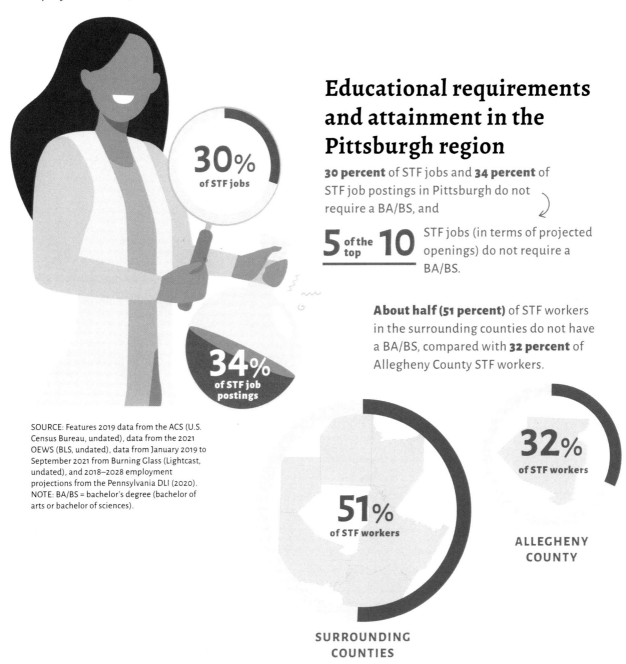

Educational requirements and attainment in the Pittsburgh region

30 percent of STF jobs and **34 percent** of STF job postings in Pittsburgh do not require a BA/BS, and

5 of the top **10** STF jobs (in terms of projected openings) do not require a BA/BS.

About half (51 percent) of STF workers in the surrounding counties do not have a BA/BS, compared with **32 percent** of Allegheny County STF workers.

30% of STF jobs

34% of STF job postings

51% of STF workers

SURROUNDING COUNTIES

32% of STF workers

ALLEGHENY COUNTY

SOURCE: Features 2019 data from the ACS (U.S. Census Bureau, undated), data from the 2021 OEWS (BLS, undated), data from January 2019 to September 2021 from Burning Glass (Lightcast, undated), and 2018–2028 employment projections from the Pennsylvania DLI (2020).
NOTE: BA/BS = bachelor's degree (bachelor of arts or bachelor of sciences).

other training opportunities, it is still not the norm for companies in the Pittsburgh region to hire individuals without bachelor's degrees for entry-level computer science jobs.

In Appendix B, we list 23 sub-baccalaureate STF occupations that pay family-sustaining wages in Pittsburgh, are not highly likely to be automated, and have at least 300 workers employed in Pittsburgh. This includes several engineering, technician, and related occupations; several health occupations; and occupations from our production and related occupations group. Some of these occupations require some postsecondary education, but nine are accessible to workers with a high school diploma. Fifteen of the 23 are also projected to increase their local employment over the next decade.

Pittsburgh Employers May Lag National Competitors in Adopting Skills-Based Hiring

Most regional STF employers represented in our focus groups reported difficulty with recruiting and retaining STF workers who meet their desired qualifications. To diversify the workforce and tap new talent pools, it has become the national norm in some STF sectors to accept lower levels of education or emerging certifications for entry-level jobs. Our data on the Pittsburgh region's educational attainment suggests that skills-based hiring, rather than degree-based hiring, may enhance the geographic diversity of the STF workforce; residents in the surrounding counties are much less likely to hold a bachelor's degree.

However, changing hiring norms can be difficult. Although many employer focus group participants reported that they are reconsidering traditional educational requirements in favor of skills-based hiring, several other stakeholders did not perceive sufficient buy-in for this kind of shift in hiring practices. The difference in local perceptions suggests that more research may be needed to understand the potential barriers to adoption of skills-based hiring.

> "[In other regions], all these big companies have signed off on, 'This [non-degree program] is enough of the skills that we need.' And it's not yet the norm in Pittsburgh. And I think if we want to elevate ourselves to be this tech mecca that we want to be—that we're purporting to be—we have to be willing to hire that entry-level talent and let there be on-ramps into it."
> —Economic development focus group participant

Accessible, Career-Relevant Education

An ideal education and training ecosystem contains multiple entry points; enables transitions between pathways; supports workers and learners with key wraparound supports and free-flowing, credible information; fosters equitable and welcoming STF workplaces; and achieves high-wage, stable employment outcomes (Batchelor et al., 2021; Zaber, Karoly, and Whipkey, 2019). To assess the accessibility and career relevance of Pittsburgh's education and training ecosystem, we draw on our scan of the region's education and training programs, our focus groups, the literature review, and an analysis of public transit offerings. Our analyses suggest that, despite a variety of education and training institutions offering relevant programs, Pittsburgh's regional pipeline for STF career-relevant education and career advancement is leaky and disjointed. We identified four regional challenges: large variation in the sufficiency of education and training providers by county and sector; the loss of qualified potential workers to other regions and countries; unclear connections between stages of education, training, and career; and significant barriers to accessing and progressing in education and training that is aligned with STF career pathways, particularly for individuals underrepresented in STF fields. We discuss each of these challenges in this section.

The Breadth of the Region's Education and Training Programs May Need to Expand in Scope and in Scale

The region is home to a large variety of STF training providers, but many economic development and employer focus group participants expressed concerns that the pipeline of training programs is insufficient in number and the diversity of skills that they offer. Economic development and employer participants perceived that training providers are stagnant in the trainings they offer and lack the insight from employers and the institutional agility to update training offerings. As a result, participants reported believing that the region's programs are not training for the "jobs of the future."[22] Employer focus group participants also expressed concern that local training capacity is insufficient to help address worker shortages in some industries, such as sub-baccalaureate nursing careers.

Our scan of the education and training landscape in the Pittsburgh region identified nearly 70 area providers (see Figure 2.5) offering postsecondary STF programs ranging from sub-baccalaureate certificates requiring only a few weeks to master's and doctoral degrees with multiyear commitments.[23] Allegheny County is home to the most providers by far across the seven counties in the Pittsburgh MSA, accounting for about 60 percent of them. However, STF programs are available at most credential levels in most counties in the region. Offerings across credential levels are most limited in Armstrong County. In addition, the region is home to 16 career and technical education centers that offer opportunities for high school students to earn industry-recognized credentials and occupational licenses in STF fields and more than 60 providers of registered apprenticeships in STF occupations.

Programs are available across STF occupational groups, with health fields being the most prevalent (number of providers) and having the greatest breadth of offerings, from short-term certificates to advanced degrees. This parallels the large share of health employment in the region. Providers of computing and math programs are the next most common, with several recently launched coding boot camps in the area. In all, more than 15,000 students completed STF programs at Pittsburgh-area institutions in the most recent year of data available from the U.S. Department of Education (2019–2020 academic year). To add context, according to data from Burning Glass, there were approximately 59,000 STF job openings in the Pittsburgh region over this period. The number of students completing STF programs should be considered a conservative estimate because it does not include those who complete workforce training programs that are not tabulated in the education data.[24] Additionally, there were more than 1,000 active apprentices in STF occupations in the region as of early 2022.[25]

---

[22] However, getting workers to train for "jobs of the future" may be difficult because some of these jobs have yet to materialize. Local employment opportunities for these jobs may be a necessary first step to create demand for these programs.

[23] Our counts of providers reflect tabulations at the institution-county level. For example, the University of Pittsburgh is counted twice, once for its main campus in Allegheny County and again for its Greensburg campus in Westmoreland County. See Appendix C for details on the methods and findings of our landscape scan.

[24] These counts also exclude completions at the four Penn State branch campuses in the Pittsburgh MSA. Those completions are not separately identified in Department of Education data and are grouped with completions at Penn State's main campus at University Park.

[25] Counts of active STF apprentices are derived from our analyses of data from the U.S. Department of Labor's Registered Apprenticeship Partners Information Database System (RAPIDS) (U.S. Department of Labor, Employment and Training

FIGURE 2.5
Training Program Distribution and Transit Accessibility

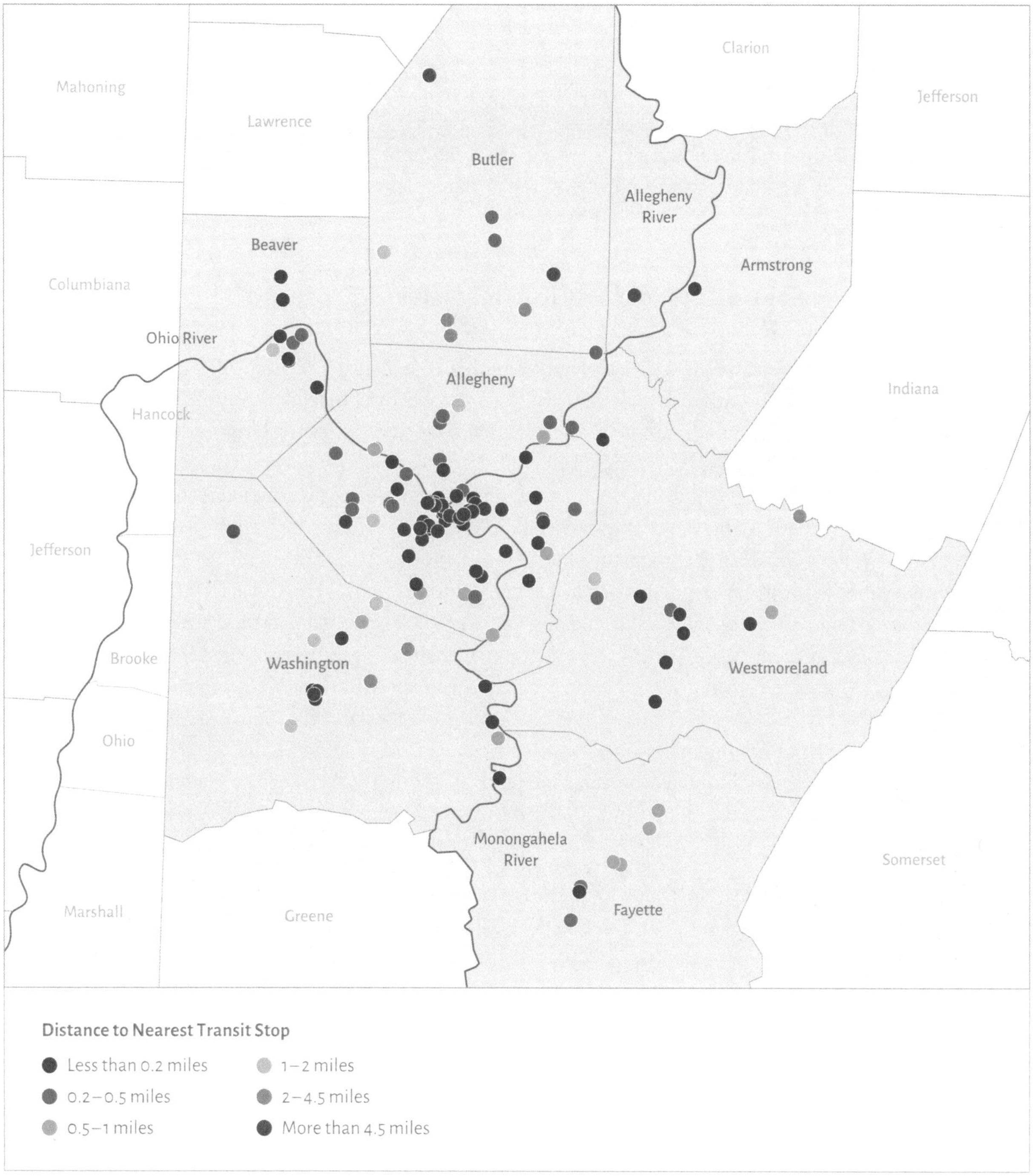

Distance to Nearest Transit Stop

- Less than 0.2 miles
- 0.2–0.5 miles
- 0.5–1 miles
- 1–2 miles
- 2–4.5 miles
- More than 4.5 miles

SOURCES: Features data from the Integrated Postsecondary Education Data System (National Center for Education Statistics, undated); data from the U.S. Department of Labor Eligible Training Provider List; and searches by the authors for institutions and data from Google Maps, county transit agencies, and the Seldin/Haring-Smith Foundation Public Transit Map for transit stops (Seldin/Haring-Smith Foundation, undated).

Administration, undated). We used a RAPIDS occupational code to O*NET-SOC code crosswalk to flag apprenticeships as STF. See Appendix C for details.

Even with sufficient regional training programs, logistical barriers, such as those related to finances, child care, and other family obligations, can stop potential learners from pursuing education and training (Deng, Zaza, and Armstrong, 2020; Henley and Roberts, 2016; Pierszalowski, Bouwma-Gearhart, and Marlow, 2021; Tsui, 2007; Wang, 2013). Several education and training providers noted that wraparound supports (e.g., financial supports, transportation, food, social services) that were offered early in the training pipeline (during K–12 education) become less common later (during postsecondary education), leading to challenges retaining learners. Transportation may be a particular challenge for STF learners and workers in surrounding counties, where both providers and students' homes may be beyond a reasonable walking (or biking) distance from public transit, and where transit schedules are much more limited and ride times are much longer (see Figure 2.5).

Enrollment at Regional Institutions May Not Guarantee an Expanded Local Workforce

Although such universities as the University of Pittsburgh and Carnegie Mellon University (CMU) graduate large numbers of students with four-year and advanced degrees in STF fields, many of those students do not stay in the region.[26] This loss of graduates may be because they find (subjectively) better jobs elsewhere. Also, the region may be more likely to lose students who relocated for school (90 percent of incoming undergraduate students at CMU are from outside Pennsylvania, including about 20 percent from outside the country).[27] By contrast, incoming students at the University of Pittsburgh, Duquesne University, Robert Morris University, and Point Park University, are mostly—but not exclusively—Pennsylvanians. Students embarking on undergraduate degree or certificate programs at local community colleges and state-affiliated institutions (such as California University of Pennsylvania and Slippery Rock University) are overwhelmingly Pennsylvania residents. Focus group participants across the three stakeholder groups perceived that students enrolled in career and technical education or boot camp programs were more likely to seek employment locally. Initial data from a graduate-tracking project (Conzelmann et al., 2022) support this perception, showing that from 2010 to 2018, the Pittsburgh region retained just 20 percent of CMU graduates and 45 percent of University of Pittsburgh (main campus) graduates, versus 83 percent of Community College of Allegheny County graduates and 79 percent of Westmoreland County Community College graduates.[28]

Connections Between Education and Training and Career Opportunities May Be Limited

Despite the variety of educational options available for different credential levels and occupations, the pipeline between education, training, and career is disconnected. Focus group participants across all three

---

[26] This loss of graduates from Pittsburgh-area universities was documented at least as far back as 2003, when graduates of three such universities in the 1990s were surveyed about career and location decisions. Researchers found that, although there was some increase in the number of graduates that stayed, the region lost a disproportionate number of graduates of color and graduates in high-tech fields (Hansen, Ban, and Huggins, 2003).

[27] The Department of Education's Integrated Postsecondary Education Data System includes data on the residency of first-time undergraduate students by state (National Center for Education Statistics, undated). These figures reflect fall 2020 enrollments. See Table C.14 for details and data for additional Pittsburgh-area institutions.

[28] In comparison, Boston retained 22 percent of Harvard University graduates, 31 percent of Massachusetts Institute of Technology graduates, 40 percent of Boston University graduates, 41 percent of Tufts University and Brandeis University graduates, 43 percent of Boston College graduates, 54 percent of Northeastern University graduates, and 76 percent of University of Massachusetts (Boston campus) graduates over the same 2010–2018 period (Conzelmann et al., 2022). There are many additional Boston-area colleges and universities not listed here.

stakeholder groups reported that, for many STF careers, it was unclear how the stages of training connect with one another and to the pathway to employment. Apart from a select number of CBOs in local neighborhoods that offer resources and connections to professional development, local training programs, and local employers (e.g., Community Forge, Homewood Children's Village),[29] there appear to be few venues for potential workers to obtain information about career pathways from trusted sources. Although many education and training providers and employers described partnerships between K–12 and CBOs as a key enabler of such connections, the programs that participants discussed primarily focus on providing exposure to STF careers.[30]

Economic development and education and training stakeholders perceive that there is a lack of awareness of local upskilling needs and opportunities throughout an STF career. There is a diversity of needs and technology is ever-changing, making it crucial that workers can continue to grow and develop in their skill sets. Several employers described investing in on-the-job training to meet emergent or highly niche skill needs rather than hiring for individuals who have those skills, in notable contrast to other employers who did not invest in on-the-job training for fear of poaching. In the words of one employer focus group participant, "we become the—and you hear the resentment in my voice—the training ground for Google, Apple, Facebook." Economic development and education and training stakeholders perceive that the advanced manufacturing sector is more adept at taking a sector-based approach to upskilling rather than relying on individual employers to do so, leaning on the coordination of employer associations and public-private partnerships, such as Catalyst Connection and the Tristate Energy and Advanced Manufacturing Consortium.

Coordination between educational institutions and STEM employers can facilitate the development of a local STEM workforce that is ready to meet employer needs (Beninson and Alper, 2021; White and Shakibnia, 2019). However, in the Pittsburgh region, focus group participants reported a lack of sustained connections between local employers and training providers that lead to employment. Multiple participants across all stakeholder groups described that both sides lack communication and act independently of one another (one employer focus group participant referred to workforce boards "training and praying" and employers "posting and praying"). Some employer participants perceived that, despite being world-renowned, many of the four-year degree–granting institutions in the region have limited connections to industry and produce fewer graduates they think of as "work-ready" compared with other regional credentialing programs. It is noteworthy that employer participants seeking to fill positions requiring very specific credentials reported more (and more sustained) partnerships with education and training programs that supply those credentials (e.g., professional residency programs, health careers) than employer participants trying to fill positions with less specific education requirements.[31] External organizations also help support specific indus-

---

[29] Community Forge is a nonprofit dedicated to "Building an equitable economy for Wilkinsburg and the Greater Pittsburgh region by creating opportunities that promote professional development, financial wellbeing, and entrepreneurial growth" (Community Forge, undated). Homewood Children's Village is a nonprofit that offers a continuum of services and support "for children and their families from cradle to career" in Pittsburgh's Homewood neighborhood (Homewood Children's Village, undated).

[30] This is not to say that building exposure is the only possible role for K–12 programming. For example, the Penn State Readiness Institute is a public-private partnership that prepares students for postsecondary education and careers through a suite of learner-centered activities. The Institute hosts a boot camp focused on building high schoolers' skills in artificial intelligence and machine learning. Appendix C contains more detail on high school career and technical education options aligned with STF careers.

[31] When very specific credentials are required for a job, programs resulting in those credentials likely convey a greater share of the totality of the knowledge and skills needed to perform the job than when educational requirements are more general. Employers, therefore, may have a greater incentive to develop close partnerships with programs that produce graduates that are essentially immediately able to perform their job functions than programs whose graduates will require considerable on-the-job training in the specifics of the role to be productive.

tries (e.g., Catalyst Connection, Black Tech Nation).[32] Apprenticeship and pre-apprenticeship models (e.g., New Century Careers)[33] were lauded by participants across stakeholder groups but were described as limited to specific industries and more common in the "traditional trades" (e.g., machinist, manufacturing technician).

An Integrated Regional Ecosystem Across Counties and Sectors

An integrated regional STF workforce ecosystem would include diverse and high-quality education and training programs that correspond with regional employment opportunities, all of which would be well distributed across the region and yet still physically accessible (via transit or remotely) to all residents. An integrated ecosystem would support STF training and employment at a variety of different credential and experience levels, including programs for high school students, early career workers, and mid-career workers who are new to STF fields or who want to further explore or advance in them. This type of system would provide a more equitable distribution of benefits from a growing STF regional economy (both geographically and across demographic groups). Currently, Pittsburgh's STF ecosystem does not exhibit all of these characteristics.

> "Really look at the region as a whole, and [do] not have Pittsburgh as . . . this hub. But look at those other places around it . . . and with some intentionality, [work] with those leaders in those counties."
> —Economic development focus group participant

---

[32] Catalyst Connection is a nonprofit providing "consulting and training services to small manufacturers in southwestern Pennsylvania" (Catalyst Connection, undated). Black Tech Nation is a multifaceted tech organization providing education, digital media, recruitment, and funding for Black technologists and entrepreneurs (Black Tech Nation, undated).

[33] New Century Careers is a nonprofit manufacturer and technical skills development organization serving southwestern Pennsylvania (New Century Careers, undated).

Geographic Distribution of Jobs and Educational Programs Requires Surrounding County Residents to Think and Act Regionally

The region's STF workforce is geographically concentrated in Allegheny, Westmoreland, and Butler Counties. The highest concentration of STF workers live in Allegheny County: Fifty-eight percent of the region's STF workers live in Allegheny County, while just 50 percent of the labor force does. Besides Allegheny, Butler County is the only county in the MSA accounting for a larger share of the MSA's STF labor force than the labor force overall (suggesting a high ratio of STF workers). However, when we look at raw numbers, Westmoreland County supplies the second-highest number of STF workers in the MSA, accounting for 12 percent of the MSA's STF labor force. More than half of employed STF workers who live in surrounding counties work outside their home county (most work in Allegheny County; see Figure 2.6), compared with about 40 percent of non-STF workers. This suggests that STF workers may be more likely than non-STF workers to seek employment regionally than within their home county; it also suggests that STF employment opportunities are largely concentrated in Allegheny County.

Similar patterns exist in the education and training landscape: STF education and training providers are concentrated in Allegheny County across all levels of credentials. For example, very few short-term non–credit-bearing programs (e.g., boot camps) are physically located in the surrounding counties (although several boot camps are currently remote). Yet community colleges in several of the surrounding counties (Beaver, Butler, and Westmoreland) offer a variety of programs and opportunities to pursue sub-baccalaureate credentials in preparation for STF jobs.

FIGURE 2.6
Where STF Workers Live and Work

58%

58 percent of STF workers from the surrounding counties work outside their home county or remotely.

22%

Just **22 percent** of STF workers from Allegheny County do so.

SOURCE: Features 2019 data from the ACS (U.S. Census Bureau, undated).

Geographic Distribution of Jobs and Educational Programs May Limit the Distribution of Benefits from the STF Ecosystem

The distribution of jobs and training programs throughout the region limits the extent to which benefits from the STF ecosystem are equitably distributed (see the box). Transportation challenges may limit access to STF training and jobs, particularly for those living outside the city of Pittsburgh.[34] Most of the economic development stakeholders who participated in our focus groups perceived minimal benefits from the STF economy for the surrounding counties, other than offering employment opportunities for some residents. However, stakeholders also believed that surrounding counties offered more physical growth opportunities for employers than Allegheny County because of federal infrastructure investment and "shovel-ready" sites.

Regional economic development planning efforts related to the STF workforce appear to be relatively rare. The stakeholders in our focus groups highlighted that leaders in surrounding counties may lack trust in regional strategic planning processes because of a historical focus of both philanthropy and economic development efforts on Allegheny County. If the region is to develop a cohesive STF strategy, future efforts might need to focus on rebuilding trust.

The Surrounding Counties Offer the STF Ecosystem Valuable Economic Resources

Pittsburgh's regional ecosystem is concentrated in Allegheny County, where most regional employment and training opportunities are located. However, the surrounding counties offer the STF ecosystem valuable economic resources. The surrounding counties' inexpensive land could offer regional employers opportunities for physical growth. Additionally, the surrounding counties are a significant source of STF labor: Forty-two percent of the region's STF workers live in the surrounding counties. The lack of employment and training opportunities in the surrounding counties could limit residents' exposure to STF careers. Regional strategies focusing on Allegheny County may be less effective at fostering sustainable and equitable economic growth than those that intentionally incorporate the assets of the surrounding counties.

BARRIERS TO PARTICIPATION

The lack of an integrated regional STF ecosystem compounds the barriers to equitable participation in the STF workforce. Transportation barriers inhibit access to training and jobs (particularly in surrounding counties), and a lack of STF jobs in some geographic and demographic communities limits early introduction to STF fields and contributes to perceptions of STF careers as out of reach. Limited information on STF career pathways and career-relevant training inhibits informed decisionmaking.

---

[34] Transportation challenges can also inhibit city residents' access to training and jobs outside the city. For example, several of the Community College of Allegheny County's programs in STF fields have been offered only at its suburban campuses. The college's new Workforce Development and Training Center, scheduled to open in 2023 at its Allegheny Campus on the North Side, will improve accessibility of these programs to city residents who rely on public transportation.

"In those surrounding counties . . . they're already shovel-ready land spaces. Everything is ready to go. And . . . a lot of the people that don't live in Pittsburgh hub are where people are coming out for these jobs. So, if we strategically start thinking about putting things in Westmoreland County, Beaver, Butler, Fayette, Green . . . they already have the commercial land and the space that they . . . need for some of these techs."
—Economic development focus group participant

Spillover Benefits

In planning and assessing regional economic development, the spillover benefits generated by economic activity can be a helpful input. *Spillover benefits* refers to when the creation of a given job benefits more than just that worker in the broader regional economy. For instance, an additional engineering job may generate demand for transportation workers, accountants, or manufacturing occupations through supply-chain relationships. The jobs created through supply-chain relationships generate additional income that supports jobs in restaurants, health care, and retail. In this section, we examine spillover effects of STF jobs in the region and draw on our focus group conversations and an analysis of economic activity conducted using regional economic data from IMPLAN and employment data from BLS.

The spillover benefits of a job are estimated by using employment multipliers. Employment multipliers show how creating employment in a particular occupation translates into broader employment changes in a regional economy (Bivens, 2019). These multipliers (1) provide insights into the local benefits from specific occupations, (2) highlight connections within a regional economy, and (3) identify occupations or groups of

---

TERMS RELATED TO SPILLOVER BENEFITS

Employment multiplier: The number of additional jobs created from one job in a specific occupation. In this report, the *total employment multiplier* refers to the sum of the jobs created through supply-chain interactions and the jobs created through income spending.

    Supply chain employment multiplier: The number of additional jobs created from one job in a specific occupation based only on business-to-business supply-chain interactions. This multiplier is commonly referred to as the *indirect multiplier*.

    Income employment multiplier: The number of additional jobs created from one job in a specific occupation based only on personal income spending. This multiplier is commonly referred to as the *induced multiplier*.

    STF sector: The top ten sectors based on sectoral share of STF workers, conditional on the sector employing at least 2,000 workers.

occupations that could play a critical role in generating inclusive regional economic growth. In this section, we calculate the employment multipliers for Pittsburgh's STF jobs and compare them with those of Boston and Nashville. Additionally, we examine the employment multipliers of Pittsburgh's STF jobs that do and do not require a bachelor's degree. Finally, we evaluate the extent to which Pittsburgh's STF jobs create employment opportunities for the region's non-STF workers.

A detailed description of our methods is available in Appendix D. However, there are several limitations of our analysis that are important to consider when interpreting our results. First, we translate regional industry activity data from IMPLAN to job counts using the National Employment Matrix from the BLS. This translation presumes that the type and number of workers used in production within an industry is similar across the United States. No Pittsburgh-specific data are available to verify this. However, if the labor inputs (number of designers, fabricators, etc.) needed to make widgets in Pittsburgh are similar to those in Peoria or San Jose, then this assumption does not threaten our analysis.

Additionally, there are other spillovers outside employment creation. Some of these spillovers are likely to be undesirable. For instance, STF jobs may create pollution (through industrial production or through worker commutes) or raise the cost of living in the area above what many non-STF workers can afford. There are also positive spillovers that we do not capture in the analysis. For example, STF jobs may increase home prices, creating wealth for existing homeowners in the region.

Finally, employment multipliers require careful interpretation. Multipliers represent the change in regional employment given a small change in the number of jobs in a specific occupation, holding constant other regional economic factors. Large changes in the number of jobs in a specific occupation may result in large changes to other regional economic factors, making multipliers a less accurate representation of the spillovers from STF jobs. As a result, the employment multipliers reported here are generally not scalable.

STF Jobs Create Additional Jobs Throughout the Regional Economy

In Pittsburgh, the average STF job creates one additional job somewhere within the region.[35] Pittsburgh's STF employment spillovers are larger than those of the comparison regions, suggesting that a larger percentage of the money earned and spent in Pittsburgh stays within the regional economy relative to Boston and Nashville. Pittsburgh's STF jobs have higher spillovers than non-STF jobs as well. The average non-STF jobs creates 0.7 additional jobs in the Pittsburgh region. This implies that the average STF job in Pittsburgh creates 30 percent more local employment than the average non-STF job.

Table 2.2 displays the average multiplier for an STF job by the occupation's educational requirement. Here, we break occupations into two groups: those that require at least a bachelor's degree and those that require less than a bachelor's degree.[36] Again, Pittsburgh STF jobs have the highest spillovers. The multiplier for STF jobs that do not require a bachelor's degree is exceptionally high in Pittsburgh relative to comparison regions, meaning that Pittsburgh's sub-baccalaureate STF jobs generate an outsized "ripple effect" in

---

[35] For ease of interpretation, the multipliers presented in this report do not include the initial direct job created and include only the indirect multiplier (jobs created through supply-chain interactions) and the induced multiplier (jobs created through consumer spending). Thus, they represent the additional jobs created.

[36] There are some occupations for which O*NET does not have educational requirement information. There are relatively few occupations that fall into this category. For instance, there are 14 BLS SOC codes in Pittsburgh that we are unable to match with O*NET educational requirements because of this missing data issue. In Pittsburgh, these occupations create 0.71 additional jobs, versus 0.53 and 0.51 additional jobs in Boston and Nashville, respectively.

TABLE 2.2

Employment Multipliers, by Region and Educational Requirements

Region	STF Jobs That Require a BA/BS	STF Jobs That Do Not Require a BA/BS
Boston	1.04	0.65
Nashville	0.98	0.82
Pittsburgh	1.01	1.04

SOURCE: Features data from IMPLAN, O*NET, BLS National Employment Matrix, and BLS OEWS (BLS, undated).

NOTE: BA/BS = bachelor of arts or bachelor of sciences degree.

employment compared with Boston and Nashville. The average sub-baccalaureate STF job in Pittsburgh creates an additional 1.04 jobs elsewhere in the region.[37]

STF Job Growth Creates Additional Jobs in Non-STF Sectors

As discussed in the sections on educational requirements and accessible, career-relevant education, there are substantial barriers on the pathway to an STF career. Nevertheless, the spillover benefits generated by STF employment may extend to non-STF workers as well. Understanding where, and to whom, these benefits accrue is critical to ensuring that future STF growth benefits residents regardless of their career path.

To assess how much non-STF workers benefit from regional STF jobs, we identified the top STF sectors based on each sector's share of STF workers, conditional on the sector employing at least 2,000 workers.[38] As shown in Table 2.3, a job in these STF sectors creates between 0.9 and 3.7 additional jobs in the Pittsburgh region. The third column depicts the multiplier received by non-STF sectors[39]—the spillover to the region that can benefit non-STF workers.

These results suggest that STF jobs in the Pittsburgh region produce sizable economic benefits for non-STF workers. For instance, 57 percent of the economic spillover from employment in Pittsburgh's hospitals sector is "received" in non-STF sectors, resulting in approximately 0.6 additional non-STF jobs. The additional jobs received in non-STF sectors are shown in the third column of the table and range from 0.5 additional jobs from the ambulatory health care sector to 2.0 additional jobs from the utilities sector.[40] This

---

[37] Occupations in the natural gas extraction, production, and distribution fields partly drive Pittsburgh's large sub-baccalaureate STF job multiplier. Specifically, the induced multiplier for these occupations tends to be large relative to other jobs in the Pittsburgh region. These jobs do not appear to pay more than other STF jobs in the Pittsburgh region, and we do not have the data necessary to understand the spending patterns of individual workers in this occupation in Pittsburgh. However, data from other regions in the United States suggest that workers in these occupations tend to have high induced multipliers. When these occupations are omitted, Pittsburgh's sub-baccalaureate STF multiplier falls to 0.95, which is still notably higher than other regions' sub-baccalaureate multipliers.

[38] See Appendix D for a detailed explanation of our methods.

[39] *Non-STF sectors* are defined as those whose STF workers make up less than 10 percent of the sector's total workforce. In Appendix D, we describe our process for identifying STF and non-STF sectors and test the sensitivity of our STF workers percentage thresholds.

[40] We did not calculate spillovers from each non-STF sector to other non-STF sectors. However, we did calculate the spillovers to non-STF sectors from four of the largest non-STF sectors (based on total employment). These sectors are food and drink services, construction, real estate, and wholesale trade. In all cases, the spillovers from these sectors to other non-STF sectors are small. The largest spillovers are from real estate, which creates 0.9 additional jobs in non-STF sectors. However, the average spillover to non-STF sectors is only 0.6.

TABLE 2.3

Economic Spillovers from STF Sectors to Non-STF Sectors, in Economic Multipliers

STF Sector	Total Additional Jobs Created	Jobs Created in Non-STF Sectors	Pittsburgh Workers
Utilities	3.7	2.0	5,362
Chemical manufacturing	2.9	1.7	4,308
Computer and electronics manufacturing	2.0	1.0	8,836
Oil and gas extraction	2.0	0.9	3,048
Transportation equipment manufacturing	1.5	0.9	2,366
Machinery manufacturing	1.3	0.7	10,496
Publishing industries	1.2	0.7	4,432
Professional scientific and technical services	1.1	0.6	84,200
Hospitals	1.1	0.6	58,115
Ambulatory health care	0.9	0.5	70,272

SOURCE: Features data from IMPLAN, Burning Glass (Lightcast, undated); and the 2021 BLS OEWS (BLS, undated).

does not guarantee that these benefits are geographically distributed. Given the concentration of STF jobs in Allegheny County, there are likely fewer STF economic spillovers that reach the surrounding counties.

Although Pittsburgh's STF sectors produce sizable spillovers to non-STF sectors, the occupations projected to grow the most in the Pittsburgh region over the next decade (many of which are health care–related occupations) produce relatively small economic spillovers in the region. The number of current workers in each sector is shown in the last column of Table 2.3 and highlights the fact that, currently, much of Pittsburgh's STF workforce is concentrated in sectors that have relatively small economic spillovers to non-STF sectors.

Focus groups with economic development stakeholders discussing the economic spillovers from STF jobs to non-STF jobs raised the question of whether new non-STF jobs pay a family-sustaining wage. The focus group participants also noted that there could be negative spillovers from the STF economy because STF companies may put upward pressure on commercial rents near new STF job centers, making it difficult for workers to live close to where they work and compounding transportation barriers. Similarly, the economic spillovers estimated in this section do not include additional negative externalities. For instance, some STF jobs may create more pollution than the typical non-STF job, particularly those in energy-related STF industries. These points of discussion indicate that more could be done, including using available policy levers to preserve affordability and prevent displacement, to minimize or prevent negative externalities.

CHAPTER 3

Planning for the Future

The previous chapter described the state of the Pittsburgh region's STF ecosystem as of spring 2022. In this chapter, we discuss how present trends are likely to have implications for the equity and sustainability of Pittsburgh's STF ecosystem in the future and raise several areas of concern. Specifically, we discuss the value of supporting nonhealth STF employment, the implications of the region's diversity challenges for the STF workforce, how the emergence of remote work may change Pittsburgh's STF ecosystem in the future, and how the skills demanded by STF jobs may evolve.

Embracing the Value of Nonhealth STF

Jobs in health care currently account for the lion's share of STF employment in the region, constituting an outsized share of Pittsburgh STF workers compared with Boston and the United States overall. The importance of the health care sector on the region's economy generally, and the STF ecosystem in particular, cannot be denied. However, such dominance by a single sector carries risks that can—and arguably should—be mitigated.

Health occupations are projected to grow faster than other STF occupational groups over the period from 2018 to 2028 in the Pittsburgh MSA, whereas nationally, health and computing and math occupations are tied for the highest projected growth rate (see Table 3.1).[1] Specific health occupations among those expected to grow the most in the Pittsburgh MSA over the coming decade include nursing occupations, medical and physician assistants, and physical therapists. Consistent with these findings, economic development focus group participants expected that most of the region's future economic growth will be driven by existing anchor institutions, many of which are in the health sector. It is no surprise that health education and training programs also dominate the local education and training landscape for STF occupations, accounting for the most providers, the most completions, and the greatest breadth of offerings across counties and at various credential levels.

Dependence on health as the main pillar of STF employment may be unwise—nationally, the health sector is very concentrated, with few employers to choose from, and our analysis suggests that Pittsburgh's health sector is more concentrated than health sectors in other regions (see Figure 3.1).

The wage "discount" on Pittsburgh labor is larger for health workers than in the STF workforce overall; health workers are paid 10 percent less than the national average (see Appendix B, Table B.16 for full comparisons).[2] Health occupations also have smaller spillovers in terms of supporting additional jobs in the region than other STF fields.

---

[1] In this section, we synthesize analyses of job postings, employment data from OEWS, and BLS's Employment Projections and Quarterly Workforce Indicators.

[2] These differences by occupational grouping may relate to the labor market for the occupation—as a place-based sector with robust local training pipelines (see the section in Chapter 2 titled "Accessible, Career-Relevant Education" for more

TABLE 3.1

Projected Growth Overall, for STF and Non-STF Occupations, and by STF Occupational Group, National versus Pittsburgh MSA, 2018 to 2028

Occupational Group	National, 2018–2028	Pittsburgh MSA, 2018–2028
Overall	5%	4%
Non-STF occupations	4%	3%
STF occupations	9%	8%
Business, management, and related	6%	7%
Computing and math	13%	8%
Architecture and engineering	4%	5%
Health	13%	10%
Production, maintenance, and related	5%	4%
Life, physical, and social sciences	6%	5%

SOURCE: Features data for 2018–2028 from the BLS occupational employment projections data and 2018–2028 long-term projections for occupational employment from the Pennsylvania DLI.

NOTE: Detailed occupations with fewer than ten workers at baseline are suppressed in the Pittsburgh MSA projections data, resulting in a sum of detailed occupational employment in 2018 that is about 38,000 (or 3 percent) lower than overall employment.

STF economic growth in sectors other than health could help balance the distribution of employment in the Pittsburgh region. The region's emerging clusters in robotics, autonomous vehicles (AVs), and advanced manufacturing could support this balancing.[3] Focus group participants anticipated growth in these sectors if the region can produce the talent to support that growth. According to one participant, "These industries are in the R&D phase, but as they expand into manufacturing, testing, [and] deployment, there will be a much greater demand."

Although the AV industry occupies a prominent place in media headlines, AV companies have accounted for only 1 percent of the STF job postings in the Pittsburgh region since 2019.[4] In addition, positions at AV companies typically require or are held by those with a bachelor's degree, which means that they have a narrower labor supply and less potential to generate individual economic mobility (relative to high-paying sub-baccalaureate STF jobs). Robotics and advanced manufacturing companies, on the other hand, tend to include a larger set of sub-baccalaureate roles, such as robotics technicians, fabricators, and computer numerical controlled machinists. Economic development efforts in these latter two sectors thus have the potential to benefit a larger variety of Pittsburgh residents. Additionally, if AV *production* takes root in Pittsburgh, complementing existing R&D efforts, that industry may follow suit. A recent report by TEConomy notes that regions that are able to build out and integrate their AV technology ecosystems—from software to hardware and to manufacturing and testing—will be poised to capture a large share of the potential trillion-dollar autonomous systems market over the next decade (Tripp et al., 2021).

---

details), health occupations may rely more on local labor markets relative to architecture and engineering occupations (4-percent discount), which may draw in workers from around the country.

[3] These three industry groups were selected by project funders as of special interest to the region.

[4] AV companies advertise for a mix of STF workers (such as engineers, computer vision specialists, and software developers) and non-STF workers (such as lawyers, marketing managers, and benefits specialists).

FIGURE 3.1

Pittsburgh Employment in Health and Autonomous
Vehicle Sectors

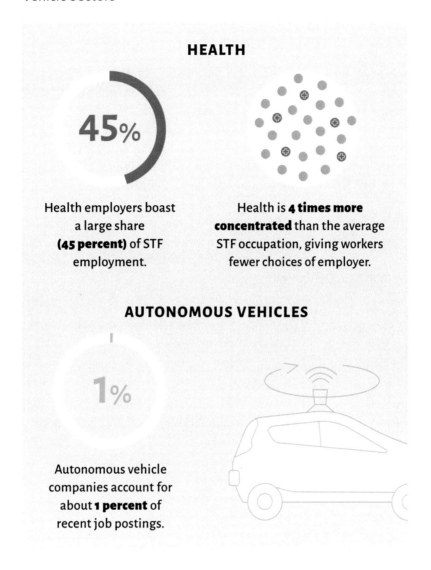

SOURCE: Features 2021 data from Burning Glass and the OEWS (Lightcast, undated; BLS, undated).

Finally, entrepreneurship is another potential avenue to diversify the sectors of Pittsburgh STF employment. Pittsburgh workers in some STF-dominant sectors are more likely to be employed at small or medium-sized firms (less than 500 employees) than their counterparts in Boston and Nashville (e.g., manufacturing firms or professional, scientific, and technical services firms).[5] However, that smaller size does not equate to newer firms—Pittsburgh workers are more likely to be employed at firms that are at least ten years old and less likely to be employed at firms less than five years old than Boston and Nashville workers. Although these

---

[5] Data on firm size and age by industry sector are from the U.S. Census Bureau, Quarterly Workforce Indicators dataset, downloaded using the Local Employment Dynamics Extraction Tool. We used a national-level industry-to-occupation matrix to identify industry sectors with higher-than-average shares of STF employment.

data are only suggestive, combining these statistics with the analysis from a prior Brookings report suggests that Pittsburgh has room to grow in terms of supporting local entrepreneurs (Andes et al., 2017).

Growing and Diversifying the STF Workforce to Sustain Ecosystem Growth

There is increasing evidence of the business case for within-firm diversity in terms of mitigating excessive risk-taking (Harjoto, Laksmana, and Yang, 2018), building effective teams, supporting additional recruitment and retention, and fueling innovation (for example, see literature reviewed in Swartz et al., 2019). Many of these dimensions will also have ripple effects by creating economic benefit for the region, which will strength the whole ecosystem, not just for its employers. The demographic data described in the next section suggest that it may be a challenge for Pittsburgh's STF employers (and the region) to reap the rewards of a diverse workforce if they hire only local talent already skilled in STF areas. Indeed, in focus groups, multiple employers and economic development stakeholders reported a lack of racial and ethnic diversity in the region as a critical barrier to hiring a diverse workforce. In this section, we discuss the current demographics of the region and barriers for attracting and retaining STF workers of color and workers from other underrepresented groups within and to the region, leveraging our analysis of focus group perceptions, data from the ACS, and our literature review.

Lack of Regional Demographic Diversity Amplifies the Importance of Making Pathways into STF Careers Accessible to All Residents

The region overall—the surrounding counties, Allegheny County, and the city of Pittsburgh—is losing residents. The region is starting from a small base of residents of color, and the city of Pittsburgh is losing Black and Asian American and Pacific Islander (AAPI) residents while growing in White non-Hispanic residents. However, the region did become slightly more racially/ethnically diverse between 2015 and 2019, with a 0.1-percent increase in the number of residents who are Black non-Hispanic (driven by Allegheny County but not the city), a 0.3-percent increase in the number of residents who are Hispanic (in the city, Allegheny County, and surrounding counties), and a 0.4-percent increase in the number of non-Hispanic AAPI residents (throughout the region, other than the city) (Table 3.2).

Increasing the diversity of Pittsburgh's STF workforce (and correspondingly, the size of the STF workforce) can come from two paths. One is to recruit Pittsburgh residents from underrepresented groups (such

TABLE 3.2

Population Change, 2015–2019

2015–2019 Change	Total	White, Non-Hispanic	Black, Non-Hispanic	Hispanic	AAPI, Non-Hispanic	Other Race, Non-Hispanic
City of Pittsburgh	−4,032	1.3%	−1.5%	0.7%	−0.5%	0.1%
Allegheny County, excluding city of Pittsburgh	−9,116	−1.7%	0.7%	0.2%	0.8%	−0.1%
Allegheny County, all	−13,148	−0.9%	0.1%	0.3%	0.5%	0.0%
Surrounding counties	−25,089	−0.6%	−0.1%	0.2%	0.3%	0.1%
Pittsburgh MSA	−38,237	−0.8%	0.1%	0.3%	0.4%	0.1%

SOURCE: Features 2015 and 2019 data from the ACS (U.S. Census Bureau, undated).

as residents of color and residents of lower socioeconomic status) to become STF workers.[6] The alternative, discussed in the section that follows, is to attract STF workers from underrepresented groups from other regions. The literature suggests that individuals from underrepresented groups may opt out of the STF workforce (consciously or unconsciously) for myriad reasons, such as a lack of exposure, role models, or support for an STF career (Blackburn, 2017; Deng, Zaza, and Armstrong, 2020); a negative perception of STF careers (Morales-Chicas et al., 2021; Shoffner et al., 2015); poor-quality K–12 and postsecondary educational experiences (Henley and Roberts, 2016; Morales-Chicas et al., 2021); lack of information about STF careers and career paths (Shoffner et al., 2015; Tsui, 2007); experiences with explicit and implicit bias in STF education and employment settings (Blackburn, 2017; Holloman et al., 2021); and insufficient resources to allow these individuals to access stable housing, quality child care, relevant educational opportunities, and reliable transportation (Chai et al., 2006; Deng, Zaza, and Armstrong, 2020; Henley and Roberts, 2016; Pierszalowski, Bouwma-Gearhart, and Marlow, 2021).

The regional importance of the factors identified in the literature was echoed by our focus group participants. For example, employers reported negative perceptions (including the history of racism and sexism in these field, and a public perception of manufacturing jobs as being "dirty") and a lack of awareness of tech jobs as barriers to hiring. Education and training providers noted that the lack of clear career pathways makes it difficult for potential workers to navigate the STF workforce system (compounding potential differences in exposure to STF careers among learners of color, rural learners, and learners from families with lower socioeconomic status), and logistical challenges, such as a lack of affordable transportation and child care, create barriers to program participation.

Just as many factors influence STF career decisionmaking, there are many strategies to better engage individuals from underrepresented groups. Two important influencers are education and training providers and employers (Blackburn, 2017; Holloman et al., 2021). For K–12 students, schools and extracurricular programs can affect the quality of an individual's foundational education and their exposure to STEM fields and STEM pathways. For postsecondary providers and for employers, targeted recruitment and advertising can increase the likelihood of reaching individuals from underrepresented groups. Organizational culture (both in terms of policies and procedures and the attitudes and values expressed at these institutions) can also influence career decisionmaking (Blackburn, 2017; Holloman et al., 2021). Barriers, facilitators, and strategies to encourage greater STF participation, particularly by members of underrepresented groups, are discussed in more detail in Appendix F.

Improving the Pittsburgh Region's Sense of Community and Inclusivity Could Attract STF Workers of Color from Other Regions

Despite this modest recent growth and the presence of potential STF workers of color enrolled at local colleges and universities (whose student bodies are generally more diverse than the region's population, as documented in Appendix C), attracting and retaining workers and families of color to the region remains a challenge (for example, see city population declines and regional lack of population growth among Black residents in Table 3.2). One challenge might be that the region has been described as inequitable.[7] As one

---

[6] The literature provides evidence for the value of many different dimensions of diversity. In this section, we focus on racial and ethnic diversity and socioeconomic status because they are the dimensions we can measure where Pittsburgh lags peer regions. Pittsburgh is on par with other regions in terms of the gender diversity of its STF workforce but is not at gender parity.

[7] The experiences with discriminatory treatment of Black women in interactions with health care institutions have been featured in reports, essays, and first-person accounts (Williamson, Howell, and Batchelor, 2017; Idia, 2020). Headlines such as "Pittsburgh is the Worst City in America for Black People" (Young, 2019) and "Pittsburgh: A 'Most Livable' City, but Not for

"One of the things that comes up a lot is specifically women of color . . . in child-bearing ages and poor health care. And that's a real problem, and it's something that has not maybe been addressed as satisfactorily as one would prefer to see in order to make this a destination. And especially if we're asking people right now . . . [to move] from Atlanta or Birmingham or somewhere, [come] to Pittsburgh, and then you see those types of statistics . . . it's very off-putting."
—Employers focus group participant

participant in the employers focus group noted, "When Pittsburgh is stated, in very widely circulated articles, to be the . . . worst place to be a Black person in America. . . . [W]e've got a lot of history to overcome in this region. . . . And so, when we talk to candidates about moving here and relocating here . . . that's a big barrier." For Pittsburgh to attract residents of color from other communities, the region might need to improve its reputation in terms of racial equity. The city and Allegheny County have both recently declared racism to be a public health crisis (Mock, 2020; Herring, 2020); if acted on, these declarations could pave the way for the region to once more call itself the home of "The Other Great Black Renaissance," as it was from the 1920s to 1950s (Whitaker, 2018).

Pittsburgh's comparatively low cost of housing is often cited as a regional asset that can be marketed to workers in other communities. Unfortunately, that asset may be less accessible to workers of color. The Black-White homeownership gap in Pittsburgh is 39 percentage points—8 percentage points larger than in Boston and 10 percentage points larger than in Nashville (Figure 3.2). A recent Pittsburgh report suggested a potential cause—between 2013 and 2020, less than half of mortgage applications submitted by Black applicants in Allegheny County ended in loans, compared with a 67-percent approval rate for White applicants (Boyle, 2022). According to the report, this disparity persists even when narrowing to low- and moderate-income applicants (44-percent approval rate for Black applicants compared with 60 percent for White applicants). Pittsburgh is not alone in this phenomenon—recent enforcement action by the U.S. Department of Justice and the Consumer Financial Protection Bureau noted lending practices consistent with redlining in the Philadelphia region (U.S. Department of Justice, 2022). The resulting settlement and the Justice Department's Combatting Redlining Initiative could catalyze extant local efforts to improve access to affordable housing (Jankiewicz, 2022).

Finally, focus group participants across stakeholder groups reported that the departure of professionals of color and the reluctance of workers of color to come to and stay in the region (which is consistent with

Black Women" (Mock, 2019) have further raised public awareness (locally and nationally) of the severe inequality by race in health, income, employment, and educational outcomes in the region. Furthermore, recent research has documented large and persistent disparities in infant mortality by race within Allegheny County (Schultz, Lovejoy, and Peet, 2022).

"Many [professionals of color] have left. And many of them have left for Atlanta. . . . They want to be able to go to a place where they feel like they can meet other young professionals there that look like them, that sort of there's a vibrant community for them; and that's not Pittsburgh, from their perspective."
—Employers focus group participant

our empirical data on more racial and ethnic diversity among college students than among longer-term residents) may stem from a lack of a sense of community in Pittsburgh. Other regions that are a hub for STF employment are notably more diverse—and larger—and can support assets and infrastructure that can complement that sense of community. This may feel like a "chicken and the egg" problem—how do you

FIGURE 3.2

Black-White Homeownership Gap, by Region, 2019

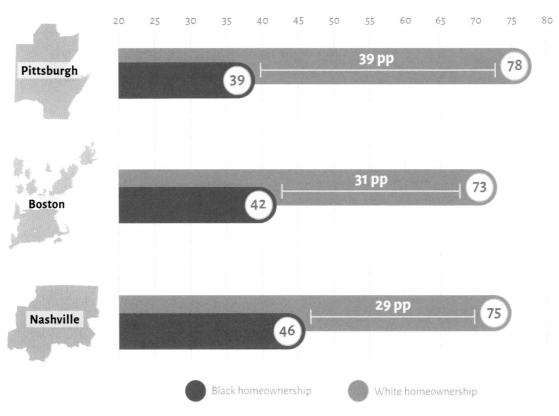

SOURCE: Features 2010 and 2019 data from the ACS (U.S. Census Bureau, undated).
NOTE: pp = percentage points.

create community if workers are unwilling to come without it? We discuss potential solutions in the Recommendations section in Chapter 4.

Balancing Regional Needs with Remote Opportunities

The coronavirus disease 2019 (COVID-19) pandemic has greatly expanded the availability of remote and hybrid options for both work and learning. For example, the number of remote job postings listed in February 2021 was more than double the January 2020 volume (Kolko, 2021). Whether this expansion results in a meaningful change in Pittsburgh remains to be seen. Pittsburgh jobs are less teleworkable relative to those in comparable cities (see Table 3.3) because of the region's higher share of health employment and technician/production employment, both of which are primarily place-based. Moreover, relative to the share of jobs that are teleworkable, comparatively fewer Pittsburgh STF workers reported teleworking during the first two years of the COVID-19 pandemic. However, there are a subset of STF employers and STF workers who could benefit from remote-work options. Because the medium- and long-term impacts of remote work remain uncertain, we illustrate several hypothetical examples describing how telework might play out in the Pittsburgh region, drawing on data from the Current Population Survey, the O*NET and OEWS (to estimate the share of teleworkable jobs), and our scan of local education and training programs.

The implications of remote work for the Pittsburgh region largely depend on where wages for remote STF workers settle in the medium and long run. Effectively, remote work "merges" STF labor markets across regions. As a result, the factors that push STF wages up in some regions and down in others dissipate, and a single STF wage emerges that all STF employers pay and all STF workers earn in the remote work market. Some recent research suggests that this is already occurring in the market for global remote work (Brinatti et al., 2021), and other research suggests that remote work is pushing down wages in the United States (Barrero et al., 2022). The new wage will fall somewhere between the highest wage and lowest wage in the current market. For instance, the remote work STF wage will not be as high as San Francisco' STF wage, but it

> "Pittsburgh has lagged behind on a pay scale. So now, it gives people who are based here with the technical skills [the] opportunity to go and command at [a] New York or at [a] San Francisco pay grade and scale up, not having to relocate . . . but at the same time, if we're thinking about more hybrid models where people are able to work from home but then also have to show up to the office, that could attract more people from a national, even international perspective, to the Pittsburgh market for job opportunity."
> —Economic development focus group participant

TABLE 3.3

Pittsburgh STF Jobs Have Low Telework Potential and Lower Telework Take-Up

	Reported Telework, STF Workers[a]		STF "Teleworkable"[b]
Year	2020	2021	2021
Boston	52.8%	40.6%	46.6%
Nashville	34.7%	26.1%	33.6%
Pittsburgh	31.3%	23.1%	36.5%

SOURCE: Features data from the Current Population Survey, merged outgoing potation groups for reported telework and from Dingel and Neiman (2020) for "teleworkable" STF jobs.

[a] These columns display the share of STF workers living in the MSA who reported teleworking in the given year. 2020 data include May to December; 2021 data include January to December.

[b] This column displays the share of MSA STF employment that is theorized to be teleworkable based on work activities and interactions.

will not be as low as STF wages in smaller cities. How Pittsburgh fares in a remote work world depends on where the remote work wage settles relative to the region's current wages.

Worker Relocation Decisions and Regional Implications Hinge on How Remote Work Is Compensated

If the remote work wage is lower than Pittsburgh's local wage, local employers could source STF workers from areas with an even lower cost of living or different community amenities. The positive impact on STF employers could translate to regional economic growth through supply-chain interactions. According to our spillover analysis, which we discussed earlier in this report and which we discuss in more detail in Appendix D, STF supply-chain multipliers tend to create additional jobs in non-STF sectors, such as wholesale trade and administrative services. The displacement of local STF workers could result in less spending on personal services in the area, driving down demand for jobs in the hospitality sector. The employment multipliers that arise through STF workers spending their income tends to create jobs in personal services and hospitality sectors.

On the other hand, if the remote work wage is higher than Pittsburgh's local wage, Pittsburgh workers could seek higher wages from companies based elsewhere without having to relocate. Additionally, STF workers from outside the region may choose to move to Pittsburgh for lower housing costs.[8] However, Pitts-

---

STF EMPLOYMENT MULTIPLIERS

The employment multipliers that arise through STF employers' supply-chain interactions tend to create jobs in such industries as wholesale trade or administrative support services.

The employment multipliers that arise through STF workers spending their income tend to create jobs in hospitality sectors, such as restaurants.

On average, the employment created through STF workers spending their income is greater than the employment created through STF employers' supply-chain interactions in the Pittsburgh region.

---

[8] Eventually, if enough workers move to Pittsburgh for the region's low housing costs, home prices will rise, making the region less affordable. This sort of general equilibrium effect is important to keep in mind when considering how future economic scenarios may play out in the Pittsburgh region.

burgh employers may be unable to attract enough affordable talent, and some may go out of business or leave the region. The loss of STF employers could result in a decline in regional jobs in supply chain–related activities.

Remote Options Make Education and Training More Accessible

Independent of the remote work scenarios discussed earlier, remote opportunities for education and training are increasing, and Pittsburgh workers may take advantage of an expanded education and training infrastructure. Massive open online courses have existed for several years, but now a broad variety of post-secondary education providers around the United States have made full credentials obtainable online, ranging from badges and certifications to advanced degrees. Within the Pittsburgh STF ecosystem, online offerings are much more common for bachelor's degrees, post-baccalaureate certificates, and advanced degrees (see Appendix C for details). Associate degrees and sub-baccalaureate certificates are sometimes offered in hybrid models but seem to have a greater reliance on in-person classroom learning and hands-on education. We consider the role of remote work in recommendations related to expanding the labor supply and crafting a regional strategy.

> "Depending on what perspective you're thinking of . . . you can buy a house here, but people, particularly diverse candidates, don't want to stay because . . . they don't see themselves in . . . Pittsburgh. . . . So, I think remote work . . . could be beneficial if you're a company, because it is cheap land, it's opportunity to grow physically. But for the worker, . . . if you are physically based in Pittsburgh, retaining that worker might be very difficult if your goal is to keep them in Pittsburgh. That person . . . may feel like working in D.C., Atlanta . . . was more attractive to them if they have that option."
> —Economic development focus group participant

Identifying Skills for the Future STF Ecosystem

STF skills will remain in demand in the region over the next decade, but new technologies, demographic change, and other factors will change the skills STF workers need. For instance, the emerging cluster of robotics, AVs, and advanced manufacturing will likely require engineers with new skill sets to work with and develop these technologies. Understanding which skill sets are projected to grow over the next decade and aligning training and educational programs to foster these skills can prevent bottlenecks in the STF workforce pipeline (see Table 3.4).

TABLE 3.4
Skills Projected to Grow and Job Postings Requesting Those Skills

Projected Growing Skill	Percentage of Current STF Postings Requesting Skill
Python	23
Occupational health and safety	21
Lifting ability	14
Machinery	13
Manual dexterity	11
Machine learning	11
Robotics	10
Scrum	9
DevOps	9
Onboarding	6
Key performance indicators	6
Digital marketing	5
Renovation	4
Data science	4
Continuous integration	4
Atlassian JIRA	4
Kubernetes	4
Artificial intelligence	4
Risk management framework	3
Flow cytometry	3
Docker software	3
Microsoft Azure	3
Spring Boot	3
Apache Kafka	3
Commercial construction	3
All other growing skills	44

SOURCE: Features data from Burning Glass Technologies (undated).

Online job postings may have predictive value, offering a nuanced picture of the skills Pittsburgh employers currently seek in STF workers and the skills that are likely to become more important for a given occupation in the future. Using job posting data from the Burning Glass Labor Insight dashboard, we collected data on the skills requested in all STF job postings between December 31, 2018, and March 22, 2022,[9] for employers hiring workers in the Pittsburgh MSA (Burning Glass, undated; Lightcast, undated). Then we filtered

---

[9] Our goal was to balance "recency" with some inclusion of pre-pandemic data.

these skills to those that are projected to grow in demand over the next few years. Burning Glass provides skill projections at the national level; these projections are based on a predictive model that incorporates historical job posting data and other web-based data sources (Lightcast, undated).[10]

Most Skills Are Projected to Have "Stable" Demand as Jobs Grow; Demand for Machine Learning and Coding Skills May Grow Within and Across Occupations

The second column of Table 3.5 displays the top five skills for each STF occupation group that Burning Glass predicts will be in greater demand in the future. Specifically, Burning Glass defines a *growing skill* as one that is projected to grow by at least 10 percent relative to other skills needed for the job. The table shows the top five growing skills based on the number of job postings that currently mention the skill. According to Burning Glass' projections, most skills across STF occupation groups are projected to have "stable" demand in the future. We present the complete list of growing skills and the top overall skills in demand in Appendix B, and we highlighted the share of current STF job postings (from January 2018 to March 2022) requesting these growing skills earlier in Table 3.4.

The skills with the most overlap across STF occupations—which are noted and shown in green in Table 3.5—are occupational health and safety, lifting ability, and Python. As shown in Table 3.4, Python is the most requested skill in STF jobs postings; 23 percent of STF postings request Python skills. Occupational health and safety skills are also highly requested (21 percent), as is lifting ability (14 percent). This interesting mix suggests that human resource skills, physical skills, and coding skills will likely play a vital role in the Pittsburgh STF ecosystem in the future. Machine learning skills and machinery skills are also growing in demand across multiple occupation groups.

TABLE 3.5

Projected Skill Growth in Pittsburgh, by STF Occupation Group

Occupation Group	Top Growing Skills	Top Overall Skills
Sciences	Occupational Health and Safety,[a] Machine Learning,[a] Python,[a] Flow Cytometery, Lifting Ability[a]	Biology, Experiments, Data Collection, Quality Assurance and Control
Production, etc.	Machinery, Lifting Ability, Occupational Health and Safety,[a] Drywall, Renovation	Repair, HVAC, Cooking, Hand Tools, Customer Service[a]
Health	Manual Dexterity, Occupational Health and Safety,[a] Behavioral Health, Lifting Ability,[a] Medical Triage	Patient Care, CPR, Treatment Planning, Life Support, Teaching
Engineering	Robotics, Python, Occupational Health and Safety,[a] Machinery, Lifting Ability[a]	AutoCAD, Project Management,[a] Scheduling,[a] Repair, Budgeting
Computing and Math	Python,[a] Scrum, DevOps, Machine Learning,[a] Kubernetes	SQL, Software Development, Java, Software Engineering, Project Management[a]
Business, etc.	Digital Marketing, Key Performance Indicators, Occupational Health and Safety,[a] Commercial Construction, Instagram	Budgeting, Project Management, Scheduling,[a] Marketing, Customer Service[a]

SOURCE: Features data from Burning Glass Technologies (undated).

NOTE: CPR = cardiopulmonary resuscitation. HVAC = heating, ventilation, and air conditioning. SQL = structured query language.

[a] Skills ranking in the top five across multiple occupation groups (also denoted by the coloration).

---

[10] The documentation for the Burning Glass skill projection model is available only to subscribers. The documentation states that Burning Glass uses a support vector machine and time series econometric methods to make skill projections. According to the documentation, the model is more than 90 percent accurate.

In addition to the cross-cutting skills, there are several highly occupation-specific skills that are projected to gain importance. For instance, Scrum—a project management tool commonly used in software development—is projected to grow in demand for computing and math occupations, while flow cytometry—a technique used to detect and measure chemical characteristic of cells or particles—is projected to grow in demand for science occupations.

Although some of these skills require broader contextual understanding (e.g., machine learning), some could easily be incorporated into a short-term education and training effort (e.g., occupational health and safety). We also caveat that this is not an exhaustive list of skills likely to be in demand in Pittsburgh in the future—some skills may not be growing in importance or demand but will remain in high demand because of employment levels (e.g., scheduling).

To address this second dimension of skill demand growth, the third column of Table 3.5 displays the top overall skills, regardless of whether these skills are projected to grow in demand across the STF occupation groups. Again, the cross-cutting skills are noted and shown in green. Here, customer service skills, project management, and scheduling skills are among the top requested skills across several occupation groups. The high demand for project management skills aligns with the perception from some employer focus group participants that Pittsburgh struggles to supply and retain managerial talent.

CHAPTER 4

Future-Proofing Pittsburgh's STF Workforce Ecosystem: Opportunities, Challenges, and Recommendations

The Pittsburgh STF ecosystem has many strengths, including strong robotics/autonomy and advanced manufacturing sectors, excellent colleges and universities, opportunities for workers without advanced degrees, outsized economic spillovers to the region at large, and increasing investment in the STF ecosystem, both from philanthropy (e.g., the Pitt BioForge facility) and from the federal government (e.g., the recent Build Back Better award). Although the region's STF ecosystem has challenges, many of the challenges identified in this report are not unique to the Pittsburgh region. For instance, a recent report on Boston's workforce in the wake of the COVID-19 pandemic highlighted substantial economic inequities by race/ethnicity, called for a shared regional strategy and increased collaboration among stakeholders, and emphasized the importance of defining career pathways (Eden, Fuller, and Lipson, 2021). Similarly, Nashville's *Vital Signs* report describes a region without a significant population decline still struggling to field enough workers, driven by lower labor force participation, demographic shifts, changing preferences over job attributes, and inaccessible or unaffordable child care (Nashville Area Chamber of Commerce, 2021).

In the section that follows, we summarize the central challenges and opportunities in the Pittsburgh STF workforce ecosystem: the structure of the Pittsburgh STF labor market, specific worker shortages, collection and dissemination of information, and regional collaboration on an STF workforce strategy.

Challenges and Opportunities

Pittsburgh's STF Labor Market May Need to Become More Competitive to Attract Enough Workers to Sustain STF Growth

Recent investments in Pittsburgh's STF ecosystem highlight the region's potential to become a national STF hub, drawing in STF workers and resources from around the country. However, to develop this type of "economic gravity," the region's STF labor market may need to become more competitive. At the core of the region's STF ecosystem is a puzzle: Why are wages comparatively low when the reported demand for STF workers is high (which is reflected in the "Sufficient Wages" and "Sufficient Labor Supply" sections of Chapter 2)? When businesses have trouble finding suitable workers, they typically respond by raising wages, and this is especially the case when they are paying below-market wages. However, wage increases that would make STF workers nationally competitive are not occurring in the Pittsburgh STF labor market. Without competitive wages, the region might be unable to attract workers and capital from more-competitive markets, contributing to the region's troubling demographic trends.

Our results suggest that a lack of competition in the STF labor market could be contributing to the region's low wages. As we discussed in the section of Chapter 3 titled "Embracing the value of nonhealth STF," health is a highly concentrated industry in general, and in Pittsburgh, the industry is even more concentrated, with few regional employers. And yet an outsized share of the region's STF employment is in health. The lack of regional competition for STF workers puts downward pressure on local wages.

However, the lack of regional competition for STF workers is not mirrored at the national level. Although the growth of remote work is a relatively new trend, many STF jobs already operated in a national labor market before the pandemic. This national market is why tech workers move from less competitive regions to technology hubs where competition for their skills is high. In other words, just because Pittsburgh-based employers do not appear to compete for the region's STF workers does not mean that employers in other markets do not. The story of Resilient Coders' departure paints the picture vividly (see the box). The region's low wages and reported norms surrounding credential-based hiring practices (as reported by focus group participants) might contribute to losing out on qualified STF workers.

Resolving the wage puzzle will likely require Pittsburgh to integrate into the broader national market for STF workers. Pittsburgh has comparatively low rates of immigration from out of state and from outside the United States, which may reduce innovation in the region. There is some suggestive evidence that Pittsburgh also operates in relative isolation from national employment trends. For example, Pittsburgh experienced a notably smaller downturn than the United States after the Great Recession and then experienced a slower recovery. This could indicate that the Pittsburgh labor market operates slightly differently than the national market does, although additional research is needed to confirm this hypothesis. In some scenarios,

---

RESILIENT CODERS

Resilient Coders, a Boston-based coding boot camp that markets itself to people of color without college degrees, brought its program to Pittsburgh in fall 2021. Pittsburgh was the third location, following Boston and Philadelphia. However, Resilient Coders is shutting down its Pittsburgh program after just more than a year; its third cohort will be its last. According to Resilient Coders staff, local stakeholders expressed support for the program. Staff reported that, although they perceived that the Pittsburgh cohorts were filled with successful, employable candidates, not a single graduate will be employed in Pittsburgh. Instead, graduates will be employed in other cities and most will relocate as part of their new jobs. Program staff shared that only one Pittsburgh-based company interviewed Resilient Coders' candidates, and their offers could not compete with what the candidates were being offered in other cities. Resilient Coders' staff told us that the program aims to support both personal and regional economic development, so they could not in good faith continue a program that pulls talented workers of color from the region.

With successes in Boston and Philadelphia, it is unclear why graduates were unable to find competitive offers locally. Program staff agreed with our focus group participants that Pittsburgh employers emphasize credentials over skills and experience, a viewpoint that disadvantages boot camp participants. It is possible that Resilient Coders' target salaries also played a role. Resilient Coders was targeting location-adjusted salaries in the mid-$90,000 range for Pittsburgh graduates, which is in line with the starting salaries received by graduates of the Philadelphia cohort (and lower than their Boston average). Our contact suspects that this relatively high starting salary may have dissuaded some potential employers in the Pittsburgh region who were reluctant to pay that much to workers without a bachelor's degree. This reflects the perspective of program staff; we were unable to speak with regional employers or program participants about Resilient Coders specifically.

remote work may force the region to integrate into national markets by providing local workers with better-paying employment opportunities without needing to move out of the region.

Pittsburgh's Relatively Inexpensive Housing Market Is Ideal for Young Workers, but the Region Is Facing Shortages of Young and Mid-Career Workers, Workers of Color, and Workers from Outside the Region to Drive Innovation and Sustain Economic Growth

Pittsburgh's stock of inexpensive housing makes it a potentially attractive location for young workers looking to build equity in the housing market, particularly as housing costs have become unaffordable in many other metro areas (Anthony, 2022). However, the Pittsburgh region faces specific shortages of young and mid-career workers and workers of color, particularly compared with other STF hubs (see the "Sufficient Labor Supply" section in Chapter 2 and the "Growing and Diversifying the STF Workforce" section in Chapter 3 for more details). Mitigating these shortages is critical for keeping the region attractive to employers—both local and in other regions—and therefore continuing to produce economic value for the region. If Pittsburgh's STF labor market is competitive, an increase in the number of workers may further decrease wages. However, if wages are low because of economic isolation, employer concentration and market power, or failure to attract employers with national footprints and wage levels, then an increase in labor supply should not depress wages and could move Pittsburgh to a higher-wage equilibrium.

The challenge of attracting and retaining workers is amplified if what we heard in our focus groups is widespread; that is, employers' reluctance to invest in internal talent pipelines out of fear of poaching. The relative newness of some STF fields in the region (which creates a limited stock of incumbent workers growing more senior), combined with a lack of professional development, might compound the region's reported challenges with attracting and retaining mid-career talent (for example, a report on the region's autonomy and robotics cluster echoed these challenges; see Tripp et al., 2021, for details).

Finally, Pittsburgh's comparatively fewer workers of color in the STF workforce likely stem from multiple causes: the region's historical and current demographics; a legacy of U.S. STF workplaces being unwelcoming or hostile to workers of color; and racial disparities in treatment by the region's core institutions, such as hospitals and banks (as documented by prior research), which reportedly dissuade workers of color outside the region from coming to Pittsburgh. The rise in remote work presents both an opportunity and a threat amid a national talent shortage, and Pittsburgh's future depends on how this plays out regionally.

Pittsburgh's Strong Education and Training Programs Have the Capacity to Produce STF Talent, but the Collection, Coordination, and Dissemination of Information on STF Career Pathways May Not Currently Provide Sufficient Insights into Job Opportunities and Career Outcomes

Pittsburgh is known for its excellent universities and is home to many sub-baccalaureate education and training programs focusing on STF skills. More data collection may be needed to effectively use the region's existing STF education and training assets. Focus group participants reported, and our landscape scan corroborated, that data collection on postsecondary education and training pathways in the Pittsburgh region is insufficient. See Appendixes C, E, and F for details. Current data systems are fragmented and typically do not track learners past separation from their institution, whether through graduation or noncompletion. Additional education and training pursued, career choices, employment stability, and wage outcomes are generally unknown. And this is even more apparent for the region's boot camps and other programs that fall outside Department of Education and Department of Labor data collections. Pennsylvania lags other states, such as Kentucky, Iowa, Texas, and West Virginia, in data linkages, in accessibility, and in developing user-friendly tools.[1] Such databases not only can identify realistic career pathways that pay family-sustaining wages (supporting CBOs working with clients and education providers working with students) but also can provide the scaffolding to evaluate education and workforce interventions with longitudinal, individual-level data.

The lack of clear pathways and timely information might carry over to upward mobility within STF careers, even within individual firms. Focus group participants perceived residents to lack understanding about how to train for STF careers, how to break into STF employment, and how to progress in an STF career. Participants noted that this lack of clarity was acutely experienced by the region's learners and workers of color, for whom information paucity might be compounded by stereotype threat and environmental cues (e.g., the region's few STF managers of color) (Scott et al., 2018).

The Seven Counties Composing the Pittsburgh MSA Have Unique and Varied Regional Assets That, When Combined, Could Spur and Sustain Additional STF Growth, but Pittsburgh Needs a Unifying Regional STF Strategy to Use These Assets Most Effectively and Enhance Ecosystem Resilience

Allegheny County is home to the majority of STF employers and workers, but 42 percent of STF workers live in the counties surrounding Allegheny. Additionally, the surrounding counties offer other important economic assets, such as shovel-ready potential developments, and could enhance the counties' contributions to the STF ecosystem. However, focus group participants felt that the surrounding counties' assets may be underused.

For instance, a regional strategy for STF growth could help distribute employment opportunities and training programs across the region, making them more accessible (via transit or remotely) to current and potential workers. At present, STF jobs and training programs are not well distributed across the region and are instead concentrated in Allegheny County (see the section titled "Accessible, Career-Relevant Education and Training" in Chapter 2). Many STF workers, particularly those who live in the counties surrounding Allegheny, work in a different county than their home county (a statistic that may only increase with the reduced commuting afforded by hybrid and remote work). However, limited public transit options in surrounding counties may present a barrier to accessing the training and employment opportunities in Allegheny County and throughout the region.

Additionally, a regional STF growth strategy could help align current and future employment needs with current and future skill development. For instance, a regional strategy would likely require that key stakeholders—employers, education and training providers, and economic development organizations, with the support of local governments and funders—coordinate to ensure that education and training opportunities are aligned (in terms of content and credential offered) with current and future workforce needs. These key stakeholders would also need to work to overcome information barriers and communicate the benefits of STF jobs, their education and skill requirements, and career growth opportunities to potential learners and workers.

---

[1] Both the Department of Education and the Department of Labor have established initiatives to support states in developing longitudinal data systems to track participants in education and training programs over time to allow analyses of the impacts of programs on individual outcomes and therefore inform decisionmaking for prospective learners and workers. For examples of promising practices across the states in collecting, linking, and publishing these data, see U.S. Department of Labor (2015), and, more recently, U.S. Department of Labor (undated). A 2021 report prepared as part of the Department of Education's Works Clearinghouse project includes recommendations related to the development of career pathways themselves (Cotner et al., 2021).

Focus group participants also reported a lack of coordination among education and training providers; between education and training providers and employers; and among economic development organizations, employers, and education and training providers in the region (see the section titled "Integrated Regional Ecosystem" in Chapter 2). Participants from all three focus group types spoke about the challenges of communicating the benefits of STF employment to potential workers. According to our focus group participants, it is unclear how the stages of training connect with one another and to the pathway to employment in many STF careers. Focus group participants worried that these programs may not be able to adapt to changing skill needs without increased coordination with employers. In addition, regional workforce development strategy is currently decentralized across four separate workforce investment boards—Partner4Work, Southwest Corner, Westmoreland Fayette, and Tri-County.

The increasing prevalence of telework and hybrid work underscores the need for a regional strategy. Recent migration from city to suburbs within Allegheny County (particularly among Black and AAPI residents) may lead STF workers to demand telecommuting options to reduce commute times. A regional strategy could leverage telework to more equitably distribute the spillover benefits from STF employment within the region (see the section in Chapter 3 titled "Balancing Regional Needs with Remote Opportunities"). Hybrid and remote workers in the surrounding counties (no longer commuting to Allegheny County) might spend more money in their home county.

The implications of remote work may spread beyond the regional economy. For instance, remote work may create a national labor market for STF labor, pulling the Pittsburgh economy out of its relative isolation. Given Pittsburgh's low STF wages compared with those of other regions, a national STF labor market would likely increase wages for local STF workers. The increased labor costs could force some Pittsburgh-based firms to shut down or move operations out of the region. If Pittsburgh-based firms want to remain competitive in a national STF labor market, increasing wages or offering more-enticing nonwage incentives may be necessary. On the other hand, the increased wages in a national labor market would likely benefit Pittsburgh's STF workers. Workers could "export" their services to other parts of the United States at a higher wage than they would receive if Pittsburgh remained isolated from the national market. These higher wages would likely bolster regional economic activity as STF workers spend their earnings locally on housing, health care, child support, or entertainment.

Recommendations

To build on the region's strengths while addressing the critical challenges of the region's insulated market conditions, uneven talent supply, inadequate timely information about STF career pathways, and an incomplete regional STF strategy, we have four main recommendations. These recommendations were developed and refined in collaboration with program officers from the four local foundations providing financial support for this project. The program officers shared their expertise on the feasibility, importance, and potential implementation of draft recommendations. We provide more detail about current local practices, best practices from models in other communities, and barriers and facilitators in Appendix G.

1. **Create market conditions that could expand the number of STF workers by**
    a. investing in entrepreneurs, startups, and capacity-constrained businesses to create competition in the STF labor market
    b. helping capacity-constrained businesses navigate the H-1B visa process
    c. creating a regional insurance pool for poaching to defray employer costs (and risks) of investing in workers

    d. creating targeted recruitment programs to provide information on STF career pathways to potential STF workers

    e. adopting skills-based hiring practices to expand the pool of applicants and help employers reduce hiring costs.

Pittsburgh is facing a shrinking and aging population and has been for decades (Morrison, 2004). For the region to continue to benefit from STF economic growth, workers must be presented with incentives to live, work, and learn in Pittsburgh. Currently, the incentives presented to the region's STF workers are lacking; Pittsburgh's STF wages are relatively low compared with those of other regions, even after adjusting for cost of living.[2] The region's low STF wages, coupled with employers' reported difficulty finding and retaining STF workers, suggest that the market for STF workers is not functioning optimally. The natural economic response to a real or perceived worker shortage is to raise the wage, so why are Pittsburgh STF wages not rising? Resolving this wage puzzle is critical to sustainably growing the region's STF workforce.[3]

There are several potential explanations for Pittsburgh's low STF wages despite strong business demand for STF workers. First, the region's hospitals employ a significant portion of its STF workforce. Specifically, 40 percent of Pittsburgh's STF workers are in health-related occupations. However, health employment is highly concentrated among a few employers in the region. In Appendix B, we calculate that Pittsburgh's health sector is 2.5 times more concentrated than the health sectors in comparator regions. Additionally, Pittsburgh's health sector is 4.5 times more concentrated than the region's computing and mathematics occupations. The lack of competition for STF workers in the region's largest STF sector is likely contributing to the region's low average STF wages.[4] Strategies that increase competition in Pittsburgh's STF sectors will likely increase STF wages, making the region more competitive for STF workers. **Investing in entrepreneurs, startups, and capacity-constrained businesses can create competition in the STF labor market. Examples could include programs designed to help capacity-constrained businesses navigate the H-1B visa process, which would expand the labor pool and reduce employer hiring costs.[5] Widespread adoption of skills-based hiring practices could also help employers reduce hiring costs by expanding the applicant pool.** Additionally, skills-based hiring could make identifying qualified workers easier, particularly if industry organizations identify relevant and acceptable sub-bachelor's qualifications.

A second potential explanation for Pittsburgh's wage puzzle is the region's employers' lack of investment in workers. Focus group participants noted hesitancy in making investments in employees for

---

[2] Some suggest that the low cost of living, coupled with the amenities of a legacy city, may be enough to attract workers. However, these factors have been in place for decades, and the region's population continues to shrink.

[3] It is important to acknowledge that if Pittsburgh's STF wages rise, additional "knock on" effects will also occur. For instance, higher wages will mean that there is more money in the local economy, which will likely increase the cost of living. Additionally, if Pittsburgh STF wages rise, some of the region's STF employers may no longer be profitable and may close. However, the loss of less productive firms may be positive for the region: The resources those firms hold would be released into the regional economy where they could be used by more-productive firms.

[4] Yeh, Macaluso, and Hershbein (2022) studied the effects of employer concentration in the manufacturing sector on wages. They found that, when relatively few manufacturing firms make up a large share of employment, wages for manufacturing workers are lower than they would be in a more competitive market.

[5] Pittsburgh lags peer regions in the share of workers who are foreign-born and in the use of H-1B visas, despite large international student populations at regional universities. Pittsburgh also has several organizations devoted to aiding immigrants and refugees in the region; expanding their capacity could help make Pittsburgh more of a destination for resettled refugees.

fear that those employees would be poached by other organizations. However, the lack of investment may keep employees from gaining the experience they need to move up within an organization, keeping them at lower-paying jobs. Solutions are needed to help reduce the perceived risk of poaching, creating incentives for employers to invest in workers and for workers to stay in the Pittsburgh region. **One potential solution is the creation of a regional insurance pool for poaching to defray the costs of training for employers. An insurance pool would be a novel market-based alternative to the use of noncompete agreements but would require buy-in from employers.** We are not aware of any existing similar programs in other regions. We discuss other potential region-wide strategies next.

A third potential explanation for the region's low wages is lack of information. Employers may need skills that workers are not currently bringing to the market and may be reluctant to increase wages if they suspect that they will be unable to find suitable candidates. On the other hand, employers may not understand the breadth of the region's training programs, causing them to pass over local qualified candidates.[6] Focus group participants expressed the sentiment that potential STF workers may not be aware of the variety of STF employers, the education and skills needed for entry into STF careers, and STF career pathways within employers or sectors. **Targeted recruitment programs can provide information on STF career pathways to potential STF workers. Additionally, improved data collection and dissemination on the region's training programs can provide potential workers and employers with information on the skills being produced in the region** (see Recommendation 3).

2. **Support and engage communities of color and other locally underrepresented groups to expand the local STF workforce and meet the ecosystem's evolving needs by**
   a. using a cohort-based model to establish scholarships for learners of color (and potentially other out-of-state learners) pursuing technical training or bachelor's degrees who commit to remain in the region for a set number of years
   b. establishing a local satellite campus of a minority-serving institution, such as a historically Black college or university
   c. building on regional momentum to shift the region's reputation on racial equity
   d. researching the rationale for leaving or not relocating to Pittsburgh among workers of color, particularly those with STF skills
   e. improving information access on STF career pathways (see Recommendation 3).

Pittsburgh's current and potential employers (and consequently, economic growth) may be constrained by the lack of demographic diversity in the region's overall and STF workforces. **Expanded demographic diversity can come from within the region, as new populations are engaged in STF, or from outside the region. To support this engagement, the region could address racial disparities (documented in prior research) in access to and treatment by public-facing systems (e.g., health care, housing, lending) that might dissuade workers of color from other communities from moving to Pittsburgh. This would build on momentum to shift the region's reputation on racial equity.** Focus group participants attributed difficulties in attracting and retaining workers of color to these disparities and their national publicity. A new economic justice–focused initiative in Boston, the New Commonwealth Racial Equity and Social Justice Fund, aims to uncover and dismantle systemic

---

[6] There is a growing movement among employers to walk back explicit degree requirements in favor of alternative credential pathways and other demonstrations of skill (Fuller, Langer, and Sigelman, 2022). However, apart from city government, most local employers have not yet joined this movement. Given the lower rate of bachelor's degrees within the Pittsburgh region (compared with Boston), qualifying potential workers based on translatable experience and skills can expand the STF labor supply.

racism within Boston through an ecosystem approach (Fuller, Langer, and Sigelman, 2022). The current efforts of this public-private partnership include supporting startup CBOs, building capacity to capitalize on other funding opportunities, and funding initiatives on health care equity and criminal justice reform (New Commonwealth Fund, 2022). If successful, these efforts could be a model for Pittsburgh to better cultivate opportunities and community for residents of color, helping the region attract the diverse talent needed to support STF ecosystem growth.

Pittsburgh could also leverage its world-class educational institutions to attract and retain future workers of color. **The region could use a cohort-based model (i.e., funding a group of students at the same stage) to establish scholarships for learners of color pursuing technical training or bachelor's degrees who commit to remain in the region for a set number of years. A cohort-based model can counter some of the challenges faced by the region's comparatively few learners and workers of color by establishing community and commonality. Alternatively, the region could explore establishing a local satellite campus of a minority-serving institution, such as a historically Black college or university.** This injection and centralization of a large number of learners of color could provide the critical mass needed to build out supportive communities.

Local funders could support research to understand the rationale among workers of color, particularly those with STF skills, for leaving or not relocating to Pittsburgh. If particular community attributes are sought, local governments and philanthropy could invest in those attributes, or market them if they already exist, to make the region more attractive to workers of color. Federal funding was recently awarded to the Southwestern Pennsylvania New Economy Collaborative under the Build Back Better competition. The region's proposal included expanded opportunities for workers from historically excluded communities (such as people of color, women, and rural residents) in autonomous industries and robotics, focusing on entrepreneurship and mentorship, both of which could help build community locally (Southwestern Pennsylvania New Economy Collaborative, 2022).

Finally, the informational improvements outlined in Recommendation 3 can help highlight the viability of and rewards from pursuing STF career pathways locally. More-complete information can help engage populations with less exposure to STF careers, such as learners of color, learners from rural communities, and learners of lower socioeconomic status.

3. **Build out regionally relevant, data-backed career pathways by**
   a. creating a state longitudinal data system that is comparable with those that exist in other states to track and communicate learner outcomes (and facilitate future research on career pathways)
   b. creating multiple entry points to career pathways and providing "sampler platter" options for lesser-known STF career fields.

Given the rapid change in the number and content of education and training programs in the region (e.g., the increasing local availability of boot camps) and the rate at which the skill requirements for jobs are changing, understanding which programs can lead to which careers is critical for the region's potential workers and learners. However, state-driven data challenges hinder the communication of career pathways and the ability to verify student outcomes. **Pennsylvania's Department of Education and DLI could collaborate to build a robust statewide longitudinal data system (SLDS) that is comparable to those that exist in other states to track and communicate learner outcomes and regional career pathways, with appropriate customization to the region's labor market and educational institutions.** These efforts would complement the infrastructure development proposed under Pennsylvania's 2019 grant award from the Department of Education's SLDS program. With additional financial support (and appropriate data use agreements), nondegree credential programs could share their student enrollment and completion data with a centralized database, such as this SLDS or the National Student Clearinghouse. These linkages could elucidate whether nondegree credential pathways are taken in tandem with traditional credentials or are an entirely alternate pathway, as well as what subsequent career options are available.

Employer input on career pathways is critical. If regional employers will not accept nondegree credentials or if they have highly specific skill needs, that input should shape and validate career pathways identified in the data. However, this is not to say that career pathways should be narrowly defined. Learners, and particularly STF learners, benefit from the opportunity to explore related career options. **Having multiple entry points to career pathways and providing "sampler platter" options for lesser-known STF career fields can help learners identify the right credential and career path for them.** This can take the form of a preprogram "bridge" course that gives learners exposure to different but related STF fields, reducing the risk of discovering a better-fitting career path while partway through a costly educational program.

Finally, simply cataloguing and publishing career pathways is insufficient. Career-changers and enterprising learners may seek out this information, but this modality is unlikely to shift the paths of potential workers who struggle with environmental cues that STF careers are not for them and other challenges around identity. However, disseminating career pathways information through trusted peers and mentors in the K–12 and postsecondary education systems and leveraging CBOs to adapt the information to individual contexts have proven to be promising in other communities (Ramirez, 2021). Coupling this information with wraparound supports makes these pathways feasible for workers facing more barriers. Focus group participants reported a lack of wraparound supports regionally past the K–12 level; the Southwestern Pennsylvania New Economy Collaborative's Build Back Better proposal specifically mentioned supporting wraparound services for postsecondary and workforce education, with close links to CBOs. This may prove to be a model for education and training programs in other sectors (Southwestern Pennsylvania New Economy Collaborative, 2022).

4. **Craft and implement a regional STF strategy by**
   a. focusing on sectors that generate more spillover for the region, such as utilities, chemical manufacturing, and information services
   b. supporting STF beyond the health sector
   c. developing capacity-conscious "toolboxes" that empower smaller or newer employers
   d. engaging trusted local leaders in the surrounding counties in regional economic development initiatives.

Working together to craft and implement a regional STF strategy would allow the region to benefit from economies of scale and enhance extant sector-based efforts by such organizations as TEAM Consortium, the Pittsburgh Robotics Network, the Pittsburgh Technology Council, and Catalyst Connection, among others. We discuss four potential components of a regional STF strategy: the diversification of STF sectors, supports for employers with lower internal capacity, a regional approach to poaching, and geographic distribution of economic development efforts.

As established previously, Pittsburgh's STF ecosystem is highly dependent on (few) health employers, yet health is one of the sectors generating the least spillover for the regional economy. **Intentional economic development and entrepreneurial focus on sectors that generate more spillover for the region, such as utilities, chemical manufacturing, and information services, can help ensure that residents benefit from STF investment regardless of their career paths.** The Southwestern Pennsylvania New Economy Collaborative, which is focused on robotics and autonomy, may prove to be an exemplar for regional cluster development (Southwestern Pennsylvania New Economy Collaborative, 2022).

Opportunities to support STF occupations beyond the health sector could also align with efforts to grow the green economy in the region. A concentrated effort to make Pittsburgh a center of green industries—coupled with revamping education and training programs to provide knowledge and skills that are in demand for green jobs, including for chemists, engineers and engineering technicians, production workers, and managers—could bolster the STF workforce and position the region to be a leader in the industries of the future, according to focus group participants. Investing in the green economy would capitalize on Pittsburgh's strong pathways for sub-baccalaureate jobs because many of these jobs do not require a four-year degree. These efforts could include building out a set of short-term training programs for aspiring green jobs workers, which is not dissimilar from the coding boot camps that have cropped up in recent years offering training for computing occupations. Such sectoral diversification also could improve the economic resilience of the region's employers and workers.

Our focus group participants indicated that there are many regional employers that *want* to be doing more to support DEI, recruit from out of state, interact with schools and CBOs, facilitate remote work, and provide competitive compensation packages to prospective workers, but they lack both internal capacity and the funds to hire an outside consultant. **Local funders could support the development of capacity-conscious "toolboxes" that empower smaller or newer employers to tackle these issues at their scale while also providing insight to more-established firms.**

Focus group participants, particularly those representing employer perspectives, suggested that regional employers do not invest as much in worker skill development as would be beneficial out of fear that that worker would be "poached" by another employer. In the same way that the unemployment insurance system spreads the cost of paying out worker unemployment benefits across firms, **a regional poaching insurance system could align incentives to support skill development.** For example, if regional employers—including frequent poachers—were compelled to pay a payroll tax into a poaching insurance system, employers whose workers were poached by another employer would receive a payout to defray the training investments made in that worker and the cost of hiring and training a new worker. Although this may not decrease worker turnover, it would allow workers to continue to seek better opportunities while encouraging employers to invest in workers once again. This approach might be best piloted in a close-knit sector with high talent demand and high potential returns from investing in workers, such as the region's robotics and autonomy cluster.

The counties surrounding Allegheny include regional assets that need to be used with intentionality. The lower cost of land sets surrounding counties up to receive investments that require larger land areas, such as AV test tracks. However, these large land developments do not necessarily create many permanent jobs for the surrounding counties, particularly relative to their footprint. Moreover, the jobs that are created may not be ones that pay family-sustaining wages (this is less true of place-based developments in the energy industry). As firms rethink the cost and value of their urban office space, the surrounding counties are a potential option for "right-sizing" expenditures on leases without shedding workers. **Economic development focus group participants described the importance of authentic engagement of trusted local leaders in the surrounding counties in planning for future development to create buy-in for a credible regional strategy.**

Limitations

This analysis has several important limitations. First, this project took place during the onset of the COVID-19 pandemic. Some of our data were collected entirely before the pandemic, and some were collected during the pandemic. The pandemic and accompanying recession were a dramatic shock to the world of work because demand for labor changed as consumers shifted to purchasing more goods and fewer services and because remote work became common in many industries and occupations. It is not yet clear whether these changes will persist in the years to come or how they are likely to play out in the Pittsburgh region in terms of employment locations, occupational concentration, and skill demands.

Second, although the Burning Glass job posting data are, to our knowledge, the most-current job posting data available, the number and distribution of job postings do not directly correspond to the number and distribution of job openings in the region (as one job posting may reflect more than one job opening). Jobs that are not advertised online are not included in our analysis.

Third, our IMPLAN spillover analysis draws on the national distribution of occupations across industries. It is possible that Pittsburgh's distribution may differ, but no data are available to verify this. However, if the labor inputs (number of designers, fabricators, etc.) needed to make widgets in Pittsburgh are similar to those in Peoria or San Jose, then this does not threaten our analysis. Additionally, the spillovers calculated in the analysis do not account for the economy-wide changes that would accompany large-scale changes in employment of a given occupation. For example, if the region were to double the size of its energy workforce because of investment in clean energy without changes in the underlying production, the spillovers from that larger workforce likely would be smaller per worker than what was estimated here.

Fourth, we were unable to look at demographic differences in regional labor market outcomes at a granular level (e.g., to compare wages among Black and White workers or male and female workers in the region within a given STF job), which means that we were not able to look at equity of labor market outcomes directly. Instead, we drew on the perceptions of the employers, workforce development experts, and training and education providers who participated in our focus groups. Our focus group participants are not a representative sample, and their perspectives may not be generalizable to stakeholders in these three groups throughout the region. In addition, we were not able to conduct a focus group with residents or STF workers. In many cases, our focus group participants shared their perceptions of what regional STF workers or residents perceive, believe, and how they behave. We report these perspectives, but note that they cannot be independently verified.

Additionally, our analysis of the STF workforce ecosystem was primarily constrained to systems and institutions operating after high school, apart from the pre-apprenticeship and career and technical education programs included in the landscape scan. Analysis of the STF career-related curricula provided in the regions' many schools, out-of-school-time providers, and other supplementary organizations was beyond the scope of this study. Nevertheless, these institutions—and the K–12 system as a whole—play an important role in developing and nurturing interest in STF careers and building STF career-aligned skills, and they merit further analysis in future work.

Finally, we were not able to answer all the questions posed by our analyses. It is unclear why Pittsburgh has wages that are consistently lower than the national average. Our analysis of ACS data suggests that the region's low wages are not driven by occupational composition or the average tenure in the workforce; in fact, the gap between Pittsburgh and national wages grows after we account for these factors. Wages are comparatively lowest (relative to U.S. averages) in the health care sector, where Pittsburgh has the most employer concentration, suggesting that market power may play a role.

We hypothesize, based on this evidence, that Pittsburgh's labor market may be less connected to the U.S. market overall, and thus less susceptible to economy-disrupting events, than other regions' markets.

One concrete example of this is that the Pittsburgh region experienced a notably smaller downturn than the United States overall following the Great Recession and then experienced a slower recovery. Pittsburgh STF employment is also concentrated in occupations that are "place-based" (e.g., health care occupations), where wages are more likely to be determined in the local labor market rather than in a national labor market. If the hypothesis that the Pittsburgh labor market is relatively isolated from the national market is correct, it is not clear how the regional economy will change in the wake of the pandemic. It is possible that increasing remote work may increase the region's connections with the national economy, which could create competition in the local STF labor market, lifting regional wages (to attract talent here) or allowing employers to hire more efficiently (from lower-cost labor markets), especially if Pittsburgh is able to attract more employers with a national footprint (and wage level) to the region. It is also possible that Pittsburgh will continue to have lower utilization of remote work and remain more of a regional labor market.

Closing Thoughts

As stated previously, the challenges that the Pittsburgh region is facing are not unique. Many midwestern communities have struggled with population decline as shifts in American manufacturing led to economic realignment. People of color are underrepresented in STF education and training programs and STF careers nationally. They also face barriers posed by environmental cues and by a lack of mentorship, access, and wraparound supports. Most of these challenges are also faced by rural residents and residents of lower socioeconomic status across the United States. Education and training providers are disconnected from employers in many communities. What is different in Pittsburgh are the magnitude of these barriers and the cost to the region of excluding *any* potential worker (inside or outside the region) from the region's STF ecosystem. In Pittsburgh, societal messages of "STF careers are not for me" are reinforced by a lack of visible leaders of color in STF fields, according to focus group participants. And failure to overcome these challenges has larger consequences in Pittsburgh as well; the region needs to grow both the size and diversity of its STF labor pool to remain competitive as an STF hub.

We recognize that there are many extant efforts to address these challenges (see Appendix G for more details), and our discussion of remaining challenges can help further support these and other efforts in the region. The Pittsburgh region also has many assets that will enable a bright economic future. What is undetermined at present is the shape of that future—how will the regional STF economy evolve, and how equitably will the benefits and costs be distributed? We believe that the four recommendations outlined in this chapter will set the region on a path to equitable and sustainable economic growth.

Implementing these recommendations will require regional collaboration and coordination among diverse organizations and stakeholders. Collecting real-time data on learner and worker outcomes, discerning career pathways from the data, and communicating identified career pathways to potential workers and learners will require connections among stakeholders throughout the STF ecosystem, including local governments, public and private employers, and CBOs. Crafting a regional strategy that brings all counties to the table will require a thoughtful and inclusive engagement approach that is sensitive to the region's historical context. These efforts will take time, commitment across leadership transitions, and trust.

Abbreviations

AAPI	Asian American and Pacific Islander
ACS	American Community Survey
AV	autonomous vehicle
BLS	U.S. Bureau of Labor Statistics
CBO	community-based organization
COVID-19	coronavirus disease 2019
DEI	diversity, equity, and inclusion
DLI	Department of Labor and Industry
K–12	kindergarten through grade 12
MSA	metropolitan statistical area
NSF	National Science Foundation
O*NET	Occupational Information Network
OEWS	Occupational Employment and Wage Statistics
PUMA	Public Use Microdata Area
RAPIDS	Registered Apprenticeship Partners Information Database System
R&D	research and development
SOC	Standard Occupational Classification
STEM	science, technology, engineering, and mathematics
STF	science- and technology-focused

References

Allegheny Conference on Community Development, *Inflection Point 2017–18: Supply, Demand and the Future of Work in the Pittsburgh Region*, 2017.

Andes, Scott, Mitch Horowitz, Ryan Helwig, and Bruce Katz, *Capturing the Next Economy: Pittsburgh's Rise as a Global Innovation City*, Brookings Institution, September 2017.

Annabi, Hala, and Sarah Lebovitz, "Improving the Retention of Women in the IT Workforce: An Investigation of Gender Diversity Interventions in the USA," *Information Systems Journal*, Vol. 28, No. 6, 2018.

Anthony, Jerry, "Housing Affordability and Economic Growth," *Housing Policy Debate*, May 2022.

Baron, James N., Michael T. Hannan, and M. Diane Burton, "Labor Pains: Change in Organizational Models and Employee Turnover in Young, High-Tech Firms," *American Journal of Sociology*, Vol. 106, No. 4, January 2001.

Barrero, Jose Maria, Nicholas Bloom, Steven J. Davis, Brent H. Meyer, and Emil Mihaylov, *The Shift to Remote Work Lessens Wage-Growth Pressures*, National Bureau of Economic Research, working paper 30197, July 2022.

Batchelor, R. L., H. Ali, K. G. Gardner-Vandy, A. U. Gold, J. A. MacKinnon, and P. M. Asher, "Reimagining STEM Workforce Development as a Braided River," Eos, April 19, 2021.

Benner, Katie, "Women in Tech Speak Frankly on Culture of Harassment," *New York Times*, June 30, 2017.

Beninson, Lida, and Joe Alper, "Meeting Regional STEMM Workforce Needs in the Wake of COVID-19: Proceedings of a Virtual Workshop Series," National Academies Press, 2021.

Bivens, Josh, *Updated Employment Multipliers for the U.S. Economy*, Economic Policy Institute, January 23, 2019.

Blackburn, Heidi, "The Status of Women in STEM in Higher Education: A Review of the Literature 2007–2017," *Science and Technology Libraries*, Vol. 36, No. 3, 2017.

Black Tech Nation, website, undated. As of November 1, 2022:
https://btn.vc/

BLS—*See* U.S. Bureau of Labor Statistics.

Boyle, John, *Taking Stock: A Decade in Decline for Black Homeownership in Pittsburgh*, Pittsburgh Community Reinvestment Group, March 2022.

Brinatti, Agostina, Alberto Cavallo, Javier Cravino, and Andres Drenik, *The International Price of Remote Work*, National Bureau of Economic Research, working paper 29437, October 2021.

Burchardi, Konrad B., Thomas Chaney, Tarek Alexander Hassan, Lisa Tarquinio, and Stephen J. Terry, *Immigration, Innovation, and Growth*, National Bureau of Economic Research, working paper 27075, May 2020.

Burkholder, Sophie, "Pittsburgh Tech Saw $10.5B Invested Over 10 Years. 'Momentum' Is Rapidly Growing," Technical.ly, webpage, March 23, 2022. As of November 1, 2022:
https://technical.ly/startups/pittsburgh-tech-investment-report-10-billion-years/

Burning Glass Technologies, "Data Skills Projections," webpage, undated. As of April 1, 2022:
https://help.burning-glass.com/en/articles/3669598-data-skills-projections

Catalyst Connection, webpage, undated. As of November 1, 2022:
https://www.catalystconnection.org/

Cech, Erin A., and Michelle V. Pham, "Queer in STEM Organizations: Workplace Disadvantages for LGBT Employees in STEM Related Federal Agencies," *Social Sciences*, Vol. 6, No. 1, 2017.

Chai, Sangmi, Sharmistha Bagchi-Sen, Rajni Goel, Raghav Rao, and Shambhu J. Upadhyaya, "A Framework for Understanding Minority Students' Cyber Security Career Interests," *Proceedings of the 12th Americas Conference on Information Systems*, Association for Information Systems, August 2006.

Chokshi, Niraj, "California Sues Tesla, Saying the Company Permitted Racial Discrimination at its Factory," *New York Times*, February 10, 2022.

Community Forge, "About: History," webpage, undated. As of November 1, 2022:
https://www.forge.community/about/

Conway, Brian, "ElevateBio Will Anchor the New Pitt BioForge Facility at Hazelwood Green," Technical.ly, webpage, August 25, 2022. As of November 1, 2022: https://technical.ly/civic-news/elevatebio-pitt-bioforge-hazelwood-green/

Conzelmann, Johnathan G., Steven W. Hemelt, Brad Hershbein, Shawn M. Martin, Andrew Simon, and Kevin M. Stange, *Grads on the Go: Measuring College-Specific Labor Markets for Graduates*, National Bureau of Economic Research, working paper 30088, May 2022.

Cotner, Hope, Debra Bragg, I-Fang Cheng, Sarah Costelloe, Brian Freeman, Grant Goold, Eric Heiser, Sebastian Lemire, Darlene G. Miller, Allan Porowski, Michelle Van Noy, and Elizabeth M. B. Yadav, *Designing and Delivering Career Pathways at Community Colleges: A Practice Guide for Educators*, National Center for Education Evaluation and Regional Assistance, Institute of Education Sciences, U.S. Department of Education, 2021.

Cox, Wendell, and Gerard Lucyshyn, *Demographia International Housing Affordability: 2022 Edition*, Urban Reform Institute and the Frontier Centre for Public Policy, 2022.

Day, Jennifer Cheeseman, and Cheridan Christnacht, "Women Hold 76% of All Health Care Jobs, Gaining in Higher-Paying Occupations," U.S. Census Bureau, webpage, August 14, 2019. As of November 1, 2022: https://www.census.gov/library/stories/2019/08/your-health-care-in-womens-hands.html

Deitrick, Sabina, and Christopher Briem, "Quality of Life and Demographic-Racial Dimensions of Differences in Most Liveable Pittsburgh," *Journal of Urban Regeneration and Renewal*, Vol. 15, No. 2, 2021.

Deng, Xuefei, Ibtissam Zaza, and Deborah J. Armstrong, "Factors Influencing IT Career Choice Behaviors of First-Generation College Students," *AMCIS 2020 Proceedings*, 2020.

Dingel, Jonathan I., and Brent Neiman, "How Many Jobs Can Be Done at Home?" *Journal of Public Economics*, Vol. 189, September 2020.

Eden, Will Dorsey, Joseph Fuller, and Rachel Lipson, *The Future of Boston's Workforce: The Path Forward from COVID-19—Findings from the Greater Boston Working Group*, The Boston Foundation and the Project on Workforce at Harvard, 2021.

Frogner, Bianca K., and Malaika Schwartz, "Examining Wage Disparities by Race and Ethnicity of Health Care Workers," *Medical Care*, Vol. 59, Suppl. 5, October 2021.

Fuller, Joseph, Christina Langer, and Matt Sigelman, "Skills-Based Hiring Is on the Rise," *Harvard Business Review*, February 11, 2022.

Funk, Cary, and Kim Parker, "Women and Men in STEM Often at Odds Over Workplace Equity," Pew Research Center, January 9, 2018.

Gibson, Campbell, *Population of the 100 Largest Cities and Other Urban Places in the United States: 1790 to 1990*, U.S. Census Bureau, June 1998.

Hansen, Susan B., Carolyn Ban, and Leonard Huggins, "Explaining the 'Brain Drain' from Older Industrial Cities: The Pittsburgh Region," *Economic Development Quarterly*, Vol. 17, No. 2, May 2003.

Harjoto, Maretno Agus, Indrarini Laksmana, and Ya-Wen Yang, "Board Diversity and Corporate Risk Taking," *SSRN*, January 24, 2018.

Henley, Lisa, and Phyllis Roberts, "Perceived Barriers to Higher Education in STEM Among Disadvantaged Rural Students: A Case Study," *Inquiry: The Journal of the Virginia Community Colleges*, Vol. 20, No. 1, 2016.

Herring, An-Li, "County Council Declares Racism a Public Health Crisis in Party-Line Vote," WESA News, May 5, 2020. As of November 1, 2022: https://www.wesa.fm/politics-government/2020-05-05/ county-council-deems-racism-a-public-health-crisis-in-party-line-vote

Holloman, T. K., J. London, W. C. Lee, C. M. Pee, C. H. Ash, and B. Watford, "Underrepresented and Overlooked: Insights from a Systematic Literature Review About Black Graduate Students in Engineering and Computer Science," *International Journal of Engineering Education*, Vol. 37, No. 2, 2021.

Homewood Children's Village, "About: Vision," webpage, undated. As of November 1, 2022: https://hcvpgh.org/about/

Howell, Junia, Sara Goodkind, Leah Jacobs, Dominique Branson, and Elizabeth Miller, *Pittsburgh's Inequality Across Gender and Race*, City of Pittsburgh's Gender Equity Commission, 2019.

Hoerr, John P., *And the Wolf Finally Came: The Decline of the American Steel Industry*, University of Pittsburgh Press, 1988.

Hunt, Jennifer, and Marjolaine Gauthier-Loiselle, "How Much Does Immigration Boost Innovation?" *American Economic Journal: Macroeconomics*, Vol. 2, No. 2, April 2010.

Idia, Terench, "'Listen to Black Women': Black Women's Health Is Intersectional Justice," Public Source, June 19, 2020.

Jankiewicz, Eric, "URA Weighs Joining Housing Authority to Aid New Homeowners," Public Source, September 8, 2022.

Johnson, Tim, "The Real Problem with Tech Professionals: High Turnover," *Forbes*, June 29, 2018.

KDKA-TV News Staff, "Study: Pittsburgh Is the No. 3 Most Livable City in the United States," CBS Pittsburgh, June 9, 2021.

Kerpen, Carrie, "Facebook Under Fire for Alleged Gender Discrimination in Job Advertisements," *Forbes*, September 9, 2021.

Kolko, Jed, "Remote Job Postings Double During Coronavirus and Keep Rising," Hiring Lab, March 16, 2021.

Langston, Abbie, Justin Scoggins, and Matthew Walsh, *Race and the Work of the Future: Advancing Workforce Equity in the United States*, PolicyLink, ERI, and Burning Glass Technologies, 2020.

Lightcast, "About: Lightcast Data," webpage, undated. As of November 1, 2022:
https://lightcast.io/about/data

Mahadeo, Jonathan, Zahra Hazari, and Geoff Potvin, "Developing a Computing Identity Framework: Understanding Computer Science and Information Technology Career Choice," *ACM Transactions on Computing Education*, Vol. 20, No. 1, 2020.

Makarem, Yasmeen, and Jia Wang, "Career Experiences of Women in Science, Technology, Engineering, and Mathematics Fields: A Systematic Literature Review," *Human Resource Development Quarterly*, Vol. 31, No. 1, 2020.

Manchester, Colleen Flaherty, "General Human Capital and Employee Mobility: How Tuition Reimbursement Increases Retention Through Sorting and Participation," *ILR Review*, Vol. 65, No. 4, 2012.

Marín-Spiotta, Erika, Rebecca T. Barnes, Asmeret A. Berhe, Meredith G. Hastings, Allison Mattheis, Blair Schneider, and Billy M. Williams, "Hostile Climates Are Barriers to Diversifying the Geosciences," *Advances in Geosciences*, Vol. 53, 2020.

Massachusetts Institute of Technology, "Living Wage Calculator," webpage, undated. As of November 1, 2022:
https://livingwage.mit.edu/

Mau, Wei-Cheng J., and Jiaqi Li, "Factors Influencing STEM Career Aspirations of Underrepresented High School Students," *Career Development Quarterly*, Vol. 66, No. 3, September 2018.

McGee, Ebony O., and Danny B. Martin, "'You Would Not Believe What I Have to Go Through to Prove My Intellectual Value!' Stereotype Management Among Academically Successful Black Mathematics and Engineering Students," *American Educational Research Journal*, Vol. 48, No. 6, December 2011.

McLean, Kate C., Isabella M. Koepf, and Jennifer P. Lilgendahl, "Identity Development and Major Choice Among Underrepresented Students Interested in STEM Majors: A Longitudinal Qualitative Analysis," *Emerging Adulthood*, Vol. 10, No. 2, 2022.

McManus, Joseph, and Joseph Mosca, "Strategies to Build Trust and Improve Employee Engagement," *International Journal of Management and Information Systems*, Vol. 19, No. 1, 2015.

Mock, Brentin, "Pittsburgh: A 'Most Livable' City, but Not for Black Women," Bloomberg, September 20, 2019.

Mock, Brentin, "How Racism Became a Public Health Crisis in Pittsburgh," Bloomberg, January 30, 2020.

Morales-Chicas, Jessica, Jenny Ortiz, Desiree M. Tanimura, and Claudia Kouyoumdjian, "Understanding Latino Boys' Motivation to Pursue STEM While Navigating School Inequalities," *Journal of Latinos and Education*, Vol. 13, 2021.

Morrison, Peter A., *A Demographic Overview of Metropolitan Pittsburgh*, RAND Corporation, IP-256-HHE/WFHF, 2004. As of June 22, 2022:
https://www.rand.org/pubs/issue_papers/IP256.html

Muñoz, José A., and Idalis Villanueva, "Latino STEM Scholars, Barriers, and Mental Health: A Review of the Literature," *Journal of Hispanic Higher Education*, Vol. 21, No. 1, 2022.

Nashville Area Chamber of Commerce, *Nashville Region's Vital Signs 2021*, 2021.

National Center for Education Statistics, "Integrated Postsecondary Education Data System," webpage, undated. As of November 1, 2022:
https://nces.ed.gov/ipeds/

National Governors Association, *Creating a More Equitable Workforce System: Opportunities for Governors and States*, August 2021.

New Century Careers, webpage, undated. As of November 1, 2022:
https://www.ncsquared.com/our-story/

New Commonwealth Fund, "New Commonwealth Fund Announces $2M in Grants to Fight Systemic Racism," press release, 2022.

Nickelsburg, Monica, "GeekWire HQ2 Revealed: We're Heading to Pittsburgh to Cover Its Post-Industrial Renaissance and Innovation Economy," webpage, December 12, 2017. As of November 1, 2022:
https://www.geekwire.com/2017/geekwire-hq2-revealed-heading-pittsburgh-cover-post-industrial-renaissance-innovation-economy/

Niebuhr, Annekatrin, "Migration and Innovation: Does Cultural Diversity Matter for Regional R&D Activity?" *Papers in Regional Science*, Vol. 89, No. 3, 2010.

Ong, Maria, Nuria Jaumot-Pascual, and Lily T. Ko, "Research Literature on Women of Color in Undergraduate Engineering Education: A Systematic Thematic Synthesis," *Journal of Engineering Education*, Vol. 109, No. 3, 2020.

Pennsylvania Department of Education, "PIMS. Postsecondary," webpage, undated. As of November 1, 2022:
https://www.education.pa.gov/DataAndReporting/PIMS/PIMSTPS/Pages/default.aspx

Pennsylvania Department of Labor and Industry, *Workforce Innovation and Opportunity Act Performance Reporting*, August 26, 2020.

Pennsylvania DLI—*See* Pennsylvania Department of Labor and Industry.

Perez-Johnson, Irma, and Harry Holzer, *The Importance of Workforce Development for a Future-Ready, Resilient, and Equitable American Economy*, American Institutes for Research, April 2021.

Pierszalowski, Sophie, Jana Bouwma-Gearhart, and Lindsay Marlow, "A Systematic Review of Barriers to Accessing Undergraduate Research for STEM Students: Problematizing Under-Researched Factors for Students of Color," *Social Sciences*, Vol. 10, No. 9, 2021.

Ramirez, Irma Y., "'I Want to Advocate for Our Kids': How Community-Based Organizations Broker College Enrollment for Underrepresented Students," *Education and Urban Society*, Vol. 53, No. 9, 2021.

Rothwell, Jonathan, *The Hidden STEM Economy*, Metropolitan Policy Program, Brookings Institution, 2013.

Schultz, Dana, Susan Lovejoy, and Evan Peet, "Tackling Persistent and Large Disparities in Birth Outcomes in Allegheny County, Pennsylvania," *Maternal and Child Health Journal*, Vol. 26, May 2022.

Scott, Allison, Freada Kapor Klein, Frieda McAlear, Alexis Martin, and Sonia Koshy, *The Leaky Tech Pipeline: A Comprehensive Framework for Understanding and Addressing the Lack of Diversity Across the Tech Ecosystem*, Kapor Center for Social Impact, 2018.

Seldin/Haring-Smith Foundation, "SHSF Public Transit Map," webpage, undated. As of November 1, 2022:
https://www.shs.foundation/shsf-transit-map

Shoffner, Marie F. Debbie Newsome, Casey A. Barrio Minton, and Carrie A. Wachter Morris, "A Qualitative Exploration of the STEM Career-Related Outcome Expectations of Young Adolescents," *Journal of Career Development*, Vol. 42, No. 2, 2015.

Shook, Ellyn, "How to Set—and Meet—Your Company's Diversity Goals," *Harvard Business Review*, June 25, 2021.

Southwestern Pennsylvania New Economy Collaborative, proposal, 2022. As of November 1, 2022:
https://eda.gov/files/arpa/build-back-better/finalists/concept-proposal-narrative/Southwestern%20PA%20New%20Economy%20Collaborative.pdf

Swartz, Talia H., Ann-Gel S. Palermo, Sandra K. Masur, and Judith A. Aberg, "The Science and Value of Diversity: Closing the Gaps in Our Understanding of Inclusion and Diversity," *Journal of Infectious Diseases*, Vol. 220, Suppl. 2, 2019.

Tascarella, Patty, "One of Pittsburgh's Newest Early Stage Funds Hasn't Cut a Local Check; Here's Why," *Pittsburgh Business Times*, November 18, 2021.

Tascarella, Patty, "Tech's Funding Conundrum," *Pittsburgh Business Times*, April 21, 2022.

Totonchi, Delaram A., Tony Perez, You-kyung Lee, Kristy A. Robinson, and Lisa Linnenbrink-Garcia, "The Role of Stereotype Threat in Ethnically Minoritized Students' Science Motivation: A Four-Year Longitudinal Study of Achievement and Persistence in STEM," *Contemporary Educational Policy*, Vol. 67, October 2021.

Tripp, Simon, Joseph Simkins, Deborah Cummings, Martin Grueber, and Dylan Yetter, *Forefront: Securing Pittsburgh's Break-Out Position in Autonomous Mobile Systems*, TEConomy Partners, LLC, September 2021.

Tsui, Lisa, "Effective Strategies to Increase Diversity in STEM Fields: A Review of the Research Literature," *Journal of Negro Education*, Vol. 76, No. 4, 2007.

U.S. Bureau of Labor Statistics, "Occupational Employment and Wage Statistics," webpage, undated. As of November 1, 2022:
https://www.bls.gov/oes/

U.S. Census Bureau, "American Community Survey Data," webpage, undated. As of November 1, 2022:
https://www.census.gov/programs-surveys/acs/data.html

U.S. Citizenship and Immigration Services, "H-1B Employer Data Hub," webpage, undated. As of November 1, 2022:
https://www.uscis.gov/tools/reports-and-studies/h-1b-employer-data-hub

U.S. Department of Justice, "Justice Department and Consumer Financial Protection Bureau Secure Agreement with Trident Mortgage Company to Resolve Lending Discrimination Claims," press release, July 27, 2022.

U.S. Department of Labor, "Workforce Data Quality Initiative (WDQI)," webpage, undated. As of July 5, 2022:
https://www.dol.gov/agencies/eta/performance/wdqi

U.S. Department of Labor, *Career Pathways Toolkit: A Guide for System Development*, 2015.

U.S. Department of Labor, Employment and Training Administration, "FY 2021 Data and Statistics: Registered Apprenticeship National Results Fiscal Year 2021: 10/01/2020 to 9/30/2021," webpage, undated. As of November 1, 2022:
https://www.dol.gov/agencies/eta/apprenticeship/about/statistics/2021

Venkatu, Guhan, *Rust and Renewal: A Pittsburgh Retrospective*, Federal Reserve Bank of Cleveland, February 2018.

Vitale, Patrick, *Nuclear Suburbs: Cold War Technoscience and the Pittsburgh Renaissance*, University of Minnesota Press, 2021.

Wakabayashi, Daisuke, "Lawsuit Accuses Google of Bias Against Black Employees," *New York Times*, March 18, 2022.

Wang, Xueli, "Why Students Choose STEM Majors: Motivation, High School Learning, and Postsecondary Context of Support," *American Educational Research Journal*, Vol. 50, No. 5, 2013.

Whitaker, Mark, *Smoketown: The Untold Story of the Other Great Black Renaissance*, Simon and Schuster, 2018.

White, Erin, and Ariana F. Shakibnia, "State of STEM: Defining the Landscape to Determine High-Impact Pathways for the Future Workforce," *Proceedings of the Interdisciplinary STEM Teaching and Learning Conference*, Vol. 3, No. 4, 2019.

Williamson, Heidi, Marcela Howell, and Michelle Batchelor, *Our Bodies, Our Lives, Our Voices: The State of Black Women and Reproductive Justice*, In Our Own Voice: National Black Women's Reproductive Justice Agenda, 2017.

Wiswall, Matthew, and Basit Zafar, "How Do College Students Respond to Public Information About Earnings?" *Journal of Human Capital*, Vol. 9, No. 2, 2015.

Yeh, Chen, Claudia Macaluso, and Brad Hershbein, "Monopsony in the U.S. Labor Market," *American Economic Review*, Vol. 112, No. 7, July 2022.

Yonemura, Rachel, and Denise Wilson, "Exploring Barriers in the Engineering Workplace: Hostile, Unsupportive, and Otherwise Chilly Conditions," *2016 ASEE Annual Conference and Exposition*, June 2016.

Young, Damon, "Pittsburgh Is the Worst City in America for Black People. Here's How It Can Get Better," *The Root*, September 18, 2019.

Zaber, Melanie A., Lynn A. Karoly, and Katie Whipkey, *Reimagining the Workforce Development and Employment System for the 21st Century and Beyond*, RAND Corporation, RR-2768-RC, 2019. As of November 1, 2022: https://www.rand.org/pubs/research_reports/RR2768.html